# Environment and Democratic Transition

# Technology, Risk, and Society
## An International Series in Risk Analysis

VOLUME 7

*The titles published in this series are listed at the end of this volume.*

# ENVIRONMENT AND DEMOCRATIC TRANSITION
## POLICY AND POLITICS IN CENTRAL AND EASTERN EUROPE

*Edited by*

**ANNA VARI**
*State University of New York, Albany and*
*Hungarian Academy of Sciences,*
*Institute for Social Conflict Research*

and

**PAL TAMAS**
*Hungarian Academy of Sciences,*
*Institute for Social Conflict Research*

**KLUWER ACADEMIC PUBLISHERS**
DORDRECHT / BOSTON / LONDON

Library of Congress Cataloging-in-Publication Data

Environment and democratic transition : policy and politics in Central
  and Eastern Europe / edited by Anna Vari and Pal Tamas.
      p.   cm. -- (Technology, risk and society ; v. 7)
  Includes index.
  ISBN 0-7923-2365-3 (HB : acid-free paper)
  1. Environmental policy--Europe, Eastern.  2. Environmental
policy--Former Soviet republics.  3. Europe, Eastern--Politics and
government--1989-  4. Former Soviet republics--Politics and
government.   I. Vári, Anna.  II. Támas, Pál.  III. Series.
HC244.Z9E518  1993
363.7'00947--dc20                                            93-17957


ISBN 0-7923-2365-3

---

Published by Kluwer Academic Publishers,
P.O. Box 17, 3300 AA Dordrecht, The Netherlands.

Kluwer Academic Publishers incorporates
the publishing programmes of
D. Reidel, Martinus Nijhoff, Dr W. Junk and MTP Press.

Sold and distributed in the U.S.A. and Canada
by Kluwer Academic Publishers,
101 Philip Drive, Norwell, MA 02061, U.S.A.

In all other countries, sold and distributed
by Kluwer Academic Publishers Group,
P.O. Box 322, 3300 AH Dordrecht, The Netherlands.

*Printed on acid-free paper*

Printed in the Netherlands

# CONTENTS

# ACKNOWLEDGEMENTS

We acknowledge with pleasure that the preparation of this book has been generously supported by the Center for Policy Research, State University of New York at Albany. We wish to thank Jeryl Mumpower, Director of the Center, for his personal support and contribution in the reviewing and editing process.

Our job of editing this book has been greatly facilitated by the efficient cooperation of the following reviewers: Ned Crosby, Jack Kartez, Joanne Linnerooth-Bayer, David McCaffrey, Stephen Stec, and Ola Svenson. In addition we would like to express our gratitude to Kathy Plunkett, Francis Richardson, Suzanne Wissel, and Lynne Womer for their collaboration in the review and preparation process.

# AUTHORS

*Richard N. L. Andrews*
  The University of North Carolina at Chapel Hill
  Chapel Hill, North Carolina, USA

*Bernd Baumgartl*
  Instituto Universitaro Europeo
  Firenze, Italy

*Boris Z. Doktorov*
  Russian Academy of Sciences - Institute of Sociology
  St. Petersburg, Russian Federation

*Boris M. Firsov*
  Russian Academy of Sciences - Institute of Sociology
  St. Petersburg, Russian Federation

*Judit Galambos*
  Regional Environmental Center for Central and Eastern
  Europe
  Budapest, Hungary

*Nikolai Genov*
  Bulgarian Academy of Sciences - Institute of Sociology
  Sofia, Bulgaria

*Kristalina Georgieva*
  World Bank
  Washington D.C., USA

*Judit Juhasz*
  Hungarian Statistical Office
  Budapest, Hungary

*Stanley J. Kabala*
  University of Pittsburgh
  Pittsburgh, Pennsylvania, USA

*Viatcheslav V. Safronov*
  Russian Academy of Sciences - Institute of Sociology
  St. Petersburg, Russian Federation

*Stephen Stec*
The Central European Environmental Law Initiative
Budapest, Hungary

*Viktoria Szirmai*
Hungarian Academy of Sciences - Institute for Social
Conflict Research
Budapest, Hungary

*Pal Tamas*
Hungarian Academy of Sciences - Institute for Social
Conflict Research
Budapest, Hungary

*Janos Tolgyesi*
ELTE University/Hungarian Academy of Sciences -
Communication Science Research Group
Budapest, Hungary

*Anna Vari*
University at Albany - State University of New York
Albany, New York, USA
and
Hungarian Academy of Sciences - Institute for Social
Conflict Research
Budapest, Hungary

*Oleg Yanitsky*
Russian Academy of Sciences - Institute for Employment
Research
Moscow, Russian Federation

Chapter 1

# ENVIRONMENTAL POLICY AND POLITICS IN CENTRAL AND EASTERN EUROPE: INTRODUCTION AND OVERVIEW

Anna Vari and Pal Tamas

The revolutions taking place in Central and Eastern European countries in 1989-1990 were followed by enthusiasm and high expectations about fast improvement in all spheres of the economy and society. Facing the realities of the severe economic, social, health, and environmental problems of the region, however, turned this enthusiasm into growing disappointment and skepticism. The assessment of the legacy of the former state-socialist rule and the analysis of opportunities of and constraints to the region's development have already started. This book contains studies which investigate the economic, legal, institutional, behavioral, social, and political aspects of the region's environmental policy. In addition to analyzing the lessons of past history, most papers focus also on the challenges, pitfalls and dilemmas that the region's policymakers and environmentalists have to face in the future.

In the first four chapters, overviews are presented about the changes in environmental policies during the period of transition in various Central and Eastern European countries. *Andrews* is focusing on the development of Czechoslovakian environmental policy between 1989-1992. The paper discusses how recent priorities -- including the democratization and decentralization of the government, the privatization of state enterprises, the transition to free market economy and free trade -- could be implemented consistently with the protection of the environment. While acknowledging the potentially positive environmental impacts of recent economic and social changes, Andrews points out several dangers of the transition, including the rapid and uncontrolled commercial exploitation of natural resources, the stimulation of private investments by loose and unenforced environmental standards, and the lack of statewide or regional environmental programs due to decentralization of governance.

*Kabala* follows the changes in environmental policy making and the emergence of the environmental protest movement in Poland from the early 1970s through the Solidarity era until the collapse of the state socialist system in 1989. Developments in environmental policy are analyzed against the backdrop of the deepening crisis of the economic and political system and the failed attempts of social and economic reform. Kabala claims that environmental policy and movements were shaped by three major factors: the extensive model of economic development focusing on rapid

1

A. Vari and P. Tamas (eds.), Environment and Democratic Transition, 1–4.

industrialization, the unbalanced competition among economic and environmental interests within the state bureaucratic setting, and the development of pluralism in a formally non-pluralist system which, by focusing on environmental issues, questioned the country's entire political system.

The paper of *Georgieva* investigates the opportunities for and constraints to environmental policy adjustment in Bulgaria during the period of transition to a market economy. After having summarized the economic, environmental, organizational, and behavioral legacy of state-socialist rule, the paper discusses the possibility for a sound legislative, regulatory, and institutional infrastructure and appropriate economic incentives for environmental protection. Georgieva's recommendations include the improvement of environmental monitoring, decision making, and control; promulgation and enforcement of implementable regulations; increase in economic efficiency; substitution of polluting processes and products with cleaner ones; clarification of property/use rights on environmental resources in privatization; and promotion of public participation in environmentally-related decisions.

*Stec* presents an overview of laws, regulations, and administrative practice related to public participation in environmental decisions in seven Central and Eastern European countries, including Albania, Bulgaria, the Czech Republic, Hungary, Poland, Rumania, and Slovakia. In addition to describing laws related to the protection of the environment, including environmental impact assessment, freedom of information and community right-to-know, the paper also investigates relevant provisions of the constitutions, public administrative laws, local self-government laws, and legislation on centralized and local permitting. Stec also reviews draft bills and other anticipated developments in environmental legislation and summarizes the most important institutions and activities of the green movements in each country.

The next four chapters discuss the emergence and development of environmental movements and conflicts in various Central and Eastern European countries. *Yanitsky* presents an analysis of the evolution of environmental initiatives and movements in Russia. His paper draws upon the analysis of the historical-cultural, macro-social, and situational contexts in which Russian urban initiatives and movements emerged. According to Yanitsky, the Russian context is characterized by state dominance, the lack of civil society, the lack of market economy and resources, the revitalization of traditional and national values, and a very limited cultivation of postmaterialist values (emphasis on quality of life and individual freedom versus consumption and social order). The paper analyzes the development of initiatives and movements within the framework of resource mobilization theory, paying special attention to comparing the opportunities of and the constraints to public participation processes in the East and the West.

*Szirmai* analyzes the evolution of the ecological-social movements in Hungary from the 1970s until the transition period following the democratic elections of 1990. Szirmai claims that the most important factors of environmental deterioration included the centralized economic and political structure and the lack of a civil society. It was first the intelligentsia who called attention to environmental damages, and this led initially to local conflicts and later to the emergence of national environmental movements. The paper discusses the complex relationships among the state, industry, and population with regard to the environmental issues, shaped by some idiosyncrasies of the country, including its economic openness, its interest in collaboration with the West, and the existence of a second economy and society.

The emergence and development of the environmental protest movement in Bulgaria, starting with the locally active environmental committees, continuing with the movement Ecoglasnost, and the formation of the Green Party, followed a scenario similar to that described in the earlier chapter. *Baumgartl* is focusing on the role served by the environmental issue as a vehicle to express political opposition to the Bulgarian state-socialist rule and on the major dilemmas of the environmental movement after the collapse of this system. Baumgartl claims that, due to the vehicular function of the green issue, many environmental activists in Bulgaria shifted from ecological to political issues. At the same time -- in contrary to its Western counterparts -- Bulgarian Greens support a liberal free-market economy.

The paper of *Galambos* investigates the international conflict on the construction of the Gabcikovo-Nagymaros dam system emerging in the late 1980s between Czechoslovakia, Hungary, and Austria. A detailed history of the controversy is followed by an analysis of the perspectives of the various stakeholder groups, including governments, water management lobbies, and environmentalists of the three countries. The paper indicates that there were serious conflicts between the above groups within each country, whereas there was some overlapping between the perspectives of the Czechoslovakian, Hungarian, and Austrian water management experts, and a certain degree of solidarity between the environmental groups of the various countries involved in the conflict. Galambos points out the key role of the changes in the political systems of the affected countries: At the beginning of the controversy, both Czechoslovakia and Hungary were socialist countries and both became pluralist democracies. Due to historical differences, however, the interpretation of the dam system became considerably different in these two countries: In Hungary it became a symbol of the old regime fought by opposition forces, whereas in Czechoslovakia it was considered as a symbol of Slovakia's national independence, power, and sovereignty. The paper also points to the lack of appropriate conflict management in the controversy and the negative consequences of this failure.

The last three chapters include papers on public attitude related to environmental problems. *Juhasz, Vari, and Tolgyesi* present a case study associated with the controversy about the siting of a low- and intermediate level radioactive waste disposal facility in Hungary in 1988-1990. A public opinion survey was conducted in 1989 in three different regions: in the area directly affected by the siting, in the area where the waste is generated, and in a neutral region. Results of the survey indicate a strong negative public attitude towards nuclear energy production and radioactive waste disposal in all three regions. The results indicate a relatively high trust in the competence of technical experts in selecting a site and monitoring the facility, low confidence in central and local governments, and -- with the exception of the area directly affected by the siting -- very little interest in public participation.

*Doktorov, Firsov, and Safronov* summarize the results of public opinion research conducted in 1989-1990 on the ecological consciousness of the Soviet population. The data reflect that the USSR public is highly concerned about the state of the local, regional, national, and global environment; and they are rather pessimistic about future outlooks as well. As a consequence of socio-economic circumstances, however, the readiness of the Russian population to make material sacrifices for protecting the environment is very low. People have very little confidence in official data and information, they have somewhat more trust in experts and environmentalists, but at present there is no mass support for the green movement.

*Genov* presents the results of public opinion surveys conducted in 1991 and 1992 in Bulgaria. The data reflect a decreasing public concern for environmental problems under the circumstances of a sharp economic decline and dramatically raising unemployment. The results also indicate a deep dissatisfaction with the institutions responsible for environmental protection and a general distrust in politicians. Some data show an increasing preference for individual initiatives and a fairly large willingness to pay a "Green tax," especially among the better educated and the young. However, these surveys show a similarity to the Russian study in that they also indicate the predominance of materialist over post-materialist value orientation.

Chapter 2

# ENVIRONMENTAL POLICY IN THE CZECH AND SLOVAK REPUBLIC

Richard N. L. Andrews

Environmental policy in any government includes not only laws and regulations that are specifically enacted to protect the environment, but the totality of government actions that in fact shape its environmental conditions. In the Czech and Slovak Republics these policies are now in a period of fundamental change, both in their own right and as a consequence of the country's larger economic and political transition. Some of these changes may cause improvements in environmental quality; others may allow continuation of old problems in the guise of a new ideology; still others may even create new environmental problems. The purpose of this assessment is to describe the pattern of environmental policy that is now developing in the Czech and Slovak Republics, and to provide tentative answers to two questions: first, how effectively are these policies dealing with both past and current environmental problems?   And second, to what extent are environmental protection considerations being woven into the new structures of economic and political decision-making -- market exchange, privatization of property, and democratization and decentralization of governance -- that are now fundamentally reshaping these countries' futures?

The answers to these questions must necessarily be tentative, because fundamental processes of change are still in progress:   at this writing, Czechoslovakia has only recently dismantled its Federal government and divided into its two constituent Republics.   This paper discusses both the Federal policies of the former Czechoslovakia and the policies of the two Republics over the years 1989-92, and their historical context.  Somewhat more detailed discussion is devoted to the policies of the Czech Republic, primarily because many Slovak policies are still in transition and fewer of them are available in English as yet.  The laws and institutions of this period, however, remain important inasmuch as they formed a common context for the environmental laws of both Republics.[1]

## Czechoslovakia as Context

The formerly Communist states of central and eastern Europe share several common characteristics:   a history of past domination by authoritarian Communist governance and centralized economic control, recent overthrow of those institutions, and rising aspirations for democratization and decentralization of governance, for private property and free markets, and for

*A. Vari and P. Tamas (eds.), Environment and Democratic Transition, 5–48.*

environmental protection and cleanup. A central challenge for each is to simultaneously build democratic and decentralized governance institutions, a healthy economy, *and* a healthy environment. How and even whether this can be done, however, is not yet clear. The responses of each country to these common challenges, moreover, are shaped as much by their individual differences of history, culture, economy and governance as by the common aspects of their recent history.

The former Czechoslovakia was a country of 127,899 km$^2$, about the size of the state of New York, with a population of about 15.5 million people. Approximately 60 percent of the land, two-thirds of the population, and 70 percent of the industry are located in the Czech Republic (Bohemia, Moravia, and a corner of Silesia), and the rest in the Slovak Republic to the east (State Commission for Science, Technology and Investments, 1990:5; Moldan and Schnoor, 1992:15). Unlike most of the other formerly Communist countries, its governance from 1968 on was shared federally between the national government and its two constituent republics.

In economic terms, Czechoslovakia was the most highly industrialized of any of the central and east European countries, dating not just to Communist policies but back to the 19th century role of the Czech lands in the Austro-Hungarian Empire. In the judgment of the World Bank, it was also one of the most successful of the centrally planned economies (World Bank, 1991). Key industries include fuels, machinery, metallurgy, chemicals and rubber, glass and ceramics, textiles, wood and paper, food, and armaments (*Czechoslovakia* 18-22). The Slovak economy, traditionally more reliant on agriculture and natural resource extraction, was also heavily industrialized during the 40 years of Communist rule, particularly to serve the growing Soviet and Central European markets for arms and other industrial products. In each Republic, nonetheless, over half the land remains agricultural and over one-third is forested. In 1937 Czechoslovakia's economy was equivalent in strength to Belgium's and 80 percent that of France, and its industrial sector was larger than Italy's; under the Communist regime, however, its economy became one of the most totally collectivized (over 98 percent), and by 1990 its economic wealth had lagged in relative terms to a level only 40 percent of its west European neighbors (Federal Committee for Environment, 1992b:28-29; World Bank, 1991:1).

## Environmental Policies in the Communist Era

The official doctrine of the Communist regimes held that environmental damage was exclusively a capitalist problem, reflecting the internal contradictions of the capitalist system. In the words of Soviet Academician Petryanov, for instance, "In a society with public ownership of the means of production, any environmental disruption will invariably be accidental" (1971, quoted in World Bank 1992:8-9). However, this doctrine ignored the fact that government responsibility for promoting the production activity created a conflict of interest within the government itself between production and protection priorities. Similar problems have been caused in the public sectors of the western market economies by government-owned enterprises, military facilities, and the operations of other government agencies. It is clearly true that the Communists' relentless pursuit of industrial and agricultural production was responsible for severe damage to public health and the environment in these countries. But the total environmental legacy of the Communist era is a more complex mixture, including also environmental problems that are common to the western market economies, as well as a legacy of unique environmental preservation opportunities that were at least inadvertent if not deliberate results of government policies during this era.

*Heavy Industry*

While Czechoslovakia had a highly industrial economy even before the Communist era, the policies of the centrally planned economy added heavy emphasis on industrial production at all costs. A basic problem caused by these policies was the economically and environmentally irrational restructuring of the overall economy. Under Communist policies, for instance, Czechoslovakia's overall economic structure was deliberately wrenched out of western trade relationships and forced to serve centrally determined objectives of the Soviet Union and the Central and Eastern European trade bloc. These objectives included even more intensive development of heavy and environmentally damaging industry, fueled by indigenous low-quality, high-sulfur brown coal (lignite) and by crude oil imported by pipeline from the Soviet Union.[2]

Within this fundamentally problematic structure, machinery, chemicals, and metallurgy alone accounted for over 68 percent of all industrial output, and industry as a whole was responsible for over 60 percent of the country's net material product (World Bank, 1991:8-11, 19-26). Moreover, production within these industries was concentrated in a small number of very large firms that dominated their sectors, creating no incentives for efficiency or innovation to keep pace with worldwide modernization of facilities and processes. As a result, the share of its products at world technical and economic levels declined from 40 percent in 1970 to 22 percent in 1980 (World Bank, 1991:9-10). Consequences of these patterns included both economic stagnation and the continued use of obsolete technologies that wasted resources and polluted the environment more than the new technologies that were being adopted in other countries.

Finally, at the level of individual enterprises, new manufacturing plants were often sited primarily for ideological reasons -- political symbolism, social engineering, or others -- rather than for any economic or environmental advantages in production. This practice led to the construction of many plants that were both economically inefficient and environmentally hazardous.

The structure of Czechoslovakia's economy was particularly damaging to its environment due to the intensity and inefficiency of its use of energy. By most estimates it used 2.5 to five times as much energy per dollar of GNP as any western country, and more energy per capita than any western country except the U.S. (and considerably more even than other formerly Communist countries, such as Poland and Hungary), though it is of course difficult to estimate what the real GNP was during the Communist era (Federal Committee for Environment, 1992b:69-70). Much of this energy was in the form of high-sulfur lignite (brown coal); most of the rest was from oil and natural gas imported from the Soviet Union (Federal Committee for Environment, 1991:43). In the year 1990, for instance, total consumption of primary energy in the Czech Republic was the equivalent of 51.7 million tons of oil; solid fuels were 65 percent of this total, and lignite was 67 percent of the solid fuels (personal communication, B. Moldan, 1/93).

## Resource Extraction

A related but distinct problem was the indiscriminate extraction of raw materials from the landscape, not only for Czechoslovakian industries (brown coal in particular) but also for direct export without valued added by processing (for instance kaolin for porcelain manufacture, and timber). Czechoslovakia had large numbers of relatively small deposits, which required specialized technical expertise to exploit efficiently; many were simply extracted and exported, or simply wasted (as for instance usable soils in coal-mine overburden), leaving their environmental impacts behind. A particularly damaging example was uranium extraction, which was conducted directly by Soviet personnel without Czechoslovak knowledge or control: these ores are found in many areas of the Czech geological massif, and were exploited in large but unknown quantities by sulfuric acid leaching, leaving a legacy of serious and continuing contamination.

## Large-Scale Agriculture

Between 1949 and 1960 the government of Czechoslovakia collectivized the agricultural economy, shifting from a pattern of small and mid-size family farms similar to other European countries to large-scale cooperatives and state farms. Agriculture is not one of the most important sectors in

Czechoslovakia in economic terms, amounting only six to ten percent of its net material product, but these policies appear to have significantly increased its environmental impacts, by substituting intensive use of mechanized equipment, chemical fertilizers, electricity, and heavy government subsidies to achieve high yields per hectare (albeit very low yields per worker, implying great inefficiency) (World Bank, 1991:12). These technologies were not well suited to the rolling hills and diversified topography of the Czechoslovakian landscape; consequences included increases in soil erosion, physical compaction of soils, indiscriminate use of pesticides, nitrate contamination from concentrated animal production, and other impacts (Hrbacek et al., 1989:138).

## Urbanization and Rural Depopulation

Czechoslovakia has long had a dense network of small settlements, over 98 percent of which even today have less than 5,000 inhabitants (Federal Committee for Environment, 1992b:23). During the Communist era the government deliberately merged many of these villages into larger administrative units, and also encouraged rural-to-urban migration into larger towns and cities. These policies were intended both for economic reasons (as a work force) and for broader social engineering goals (to dilute or proletarianize the "cultural elite" in the cities). One result of these policies was a significant increase in urban population densities -- as of 1987 68 percent of the population lived in towns or cities -- with concomitant environmental impacts, including not only urban air and water pollution but also housing and urban service needs, rapid expansion of cities, and others. A particularly obvious legacy was the construction of huge concrete-slab apartment cities ("block flats") in which many people now live, widely disliked for their ugliness, their lack of neighborhood commercial services, their poor energy efficiency and low-quality construction, and other characteristics. Investments in modernization of urban infrastructure, such as wastewater treatment plants and air pollution control for household heating, have not come close to keeping pace with the rate of urbanization. Hrbacek et al. (1989:139, 149) note that while the first program for construction of sewage plants was initiated in 1958, the cost of constructing wastewater facilities was not an obligatory part of the budget approval process for new housing facilities, but was allocated separately from an "ecological investments" category which had low priority.

A parallel result was significant depopulation of rural regions: the number of villages was reduced from almost 12,000 in 1961 to less than 7,000 in 1988 (State Commission for Science, Technology and Investments, 1990b:7). In the Czech Republic this process was preceded by postwar expatriation of over three million citizens of German descent, virtually one-third of the population of the Czech Republic, which had already emptied sizable areas of the countryside. This process also undermined the social and economic viability of rural villages, with a mixture of environmental impacts: more severe damage from large-scale agriculture and from expansion of heavy industry in some regions, but also diminished human impacts in other areas. Many former towns are now occupied primarily as weekend and vacation homes of city dwellers, and lack the jobs, resident populations, and commercial services to be economically viable otherwise. In some cases, especially the border areas of the former Iron Curtain, the near-total degree of rural depopulation even amounted to the creation of *de facto* natural areas where for 40 years almost no human settlement other than occasional military facilities was permitted.

Finally, municipalities under Communism were important causes of environmental problems, chiefly because they were responsible for many municipal services but had neither the resources nor the political power or will to prevent their impacts. One widely visible example was the construction of huge concrete-slab apartment cities ("block flats") in which many people now live, widely disliked for their ugliness, low-quality construction, poor energy efficiency, lack of neighborhood commercial services, and other characteristics. A second example was municipal energy production for central heating, frequently by low-quality, high-sulfur brown coal (lignite) without air pollution controls: these plants were among the most significant sources of urban air pollution, as well as those posing the most direct health hazards to population concentrations. A third example was wastewater management: some 2,500 municipalities serving 2.5 million people, for instance, had no wastewater treatment facilities, and heavy metals exceeded drinking water standards in 123 towns serving over half a million people (Moldan and Schnoor, 1992:16).[3] A fourth example was the almost universal decay of urban buildings and infrastructure, owing both to lack of budgets for maintenance and to the system of state ownership of all property: there were no private resources or ownership incentives for maintaining and modernizing most buildings and facilities, and within the centralized state budget system these needs generally received low priority.[4] More generally, investments in modernization of urban infrastructure did not come close to keeping pace with the rate of urbanization, with only the partial exception of relatively convenient public transport systems.

### Protected Areas

Czechoslovakia also instituted explicit legal protection for landscapes and land areas that had special ecological or other natural values. These areas included one national park and twenty landscape-protected formations in the Czech Republic (two of them recognized by UNESCO as biosphere reserves), and three national parks and eleven landscape-protected areas (one a UNESCO reserve) in the Slovak Republic. Together these areas amounted to approximately 13 percent of the total lands of each Republic, not including nearly 1,000 smaller sites protected for endangered species. Protected status in fact provided only nominal protection, however, and did not prevent actual damage both from deposition of pollution and from other human intrusions (Hrbacek et al., 1989:150).

## De-Emphasis on Consumer Goods

Another important policy pattern was the relative neglect of the consumer goods and services sectors during the Communist era (World Bank, 1991:9). Czechoslovakia's per capita income and consumption rose slowly but steadily throughout most of the Communist era, but both the production and the quality of consumer products lagged; access to consumer credit was limited; and commerce was largely limited to government-owned shops.

These patterns had ambiguous environmental consequences. On the one hand, the absence of a market economy restrained many of the wasteful practices that are the sources of some of the western nations' current environmental problems, such as excess packaging and throw-away goods in general (residents of the former East Germany, for instance, have commented on how much their own household trash increased almost immediately after unification). It also limited the large-scale patterns of environmental impact associated with widespread automobile usage: as of 1988 Czechoslovakia had only 187 private cars per thousand citizens (World Bank, 1991:121), but a relatively comprehensive system of public transport.

The same pattern of policies, however, as well as limited access to consumer credit, also promoted the continued use of antiquated and environmentally wasteful products, such as decades-old refrigerators and other energy-consuming appliances; and even many new products were of relatively poor quality and manufactured in resource-wasteful and environmentally damaging ways. A key challenge for environmental policy in the Czech and Slovak Republics today, in fact, is not to stop economic production and growth but to promote those elements of it that will help to modernize and replace the environmentally-damaging products and processes that are now being used.

## Pollution

The severity of environmental pollution in Czechoslovakia was identified as early as the 1960s by outside observers, and even by scientists and government reports within the country, but its full magnitude was not officially acknowledged until after the "velvet revolution" of 1989.

In 1984, for example, an article by a British scholar documented policy debates about pollution in Prague dating as far back as the beginning of the 19th century, and described in detail both the pollution levels and the policy responses to them that occurred from the 1960s on (Carter, 1984). Hrbacek et al. (1989:138) also cited scientific papers documenting industrial pollution problems in 1915, 1924, 1937, and 1940. Serious increases in air pollution began in 1945, with the decision to rapidly expand heavy industry using indigenous soft brown coal (with sulfur content frequently in excess of three percent) as fuel. From 1945 to at least 1985, Czechoslovakia's annual consumption of coal climbed steadily, from 14 million to a peak of 130 million tons per year. Papers by Czech authors as early as 1964 documented a 12-fold increase in dust fallout over one district of Prague from 1925 to 1964 (from 90 to 1,106 tons per km$^2$ per year); SO$_2$ emissions increased almost six-fold from 1959 to 1968 (Moldan and Schnoor, 1992:15-16).

Data on air quality were systematically collected beginning in 1954, and a decree setting maximum concentrations was enacted in 1966; a contour map was published by the government hygiene station in Prague in 1966, showing dust concentrations for 1965 (over 700 tons per km$^2$ in peak areas, and an average of 461 tons city-wide) (cited in Carter, 1984). A detailed study of environmental conditions in Prague was conducted in the 1970s, and many data on environmental conditions and public health were collected routinely by district hygiene stations and other government agencies, but most were declared official secrets and suppressed. The Ecological Section of the Biological Society attached to the Czechoslovak Academy of Sciences produced scientific reports on the severity of environmental conditions both in 1983 and again in 1989, for instance, but neither was allowed to be officially published (World Bank, 1992:10). To be sure, most of these data addressed only conventional emissions, and did not include information on other toxic pollutants; but enough was known to document severe problems.

With the steady expansion of brown coal use, annual emissions of SO$_2$ grew from under one million tons in 1950 to a peak of almost 3.2 million tons in 1985 (Federal Committee for Environment, 1992). As of 1988 Czechoslovakia's emissions ranked second highest in Europe, after East Germany; and releases of organic pollutants to watercourses increased steadily from municipal sources, even though industrial sources reportedly held steady or were slightly reduced during the 1980s. Average annual concentrations of SO$_2$ in Prague and several other districts was in excess of 100 ug/m$^3$, compared to both a national and a World Health Organization standard of 50 (Federal Committee for Environment, 1992:19); extreme readings of over three thousand ug/m$^3$ were recorded for 24-hour periods in Prague and other locations during atmospheric inversions (Czech Ministry, 1990; Moldan and Schnoor, 1992:15; World Bank, 1992:8). Incidence of respiratory disease was high, and correlated closely with areas and seasons of high air pollution; life expectancy was also measurably shorter in the most polluted industrial regions; and the incidence of allergies appeared to have increased six-fold (ten-fold for children) in the past ten years (Federal Committee for Environment, 1992b:81, 93-115). The average life-span in Czechoslovakia was three to six years shorter than those of most west European countries (Federal Committee for Environment, 1992c:54). Seventy percent of surface waters in the country were also considered heavily polluted, and 30 percent were so polluted that they could not sustain fish populations; 2,500 municipalities serving 2.5 million people had no wastewater treatment facilities, and heavy metals exceeded drinking water standards in 123 towns serving over half a million people (Moldan and Schnoor, 1992:16). Over 70 percent of the forests in Czechoslovakia showed some signs of damage, and over five percent were estimated dead or dying; 50 to 90 percent of many species of wildlife were considered endangered, far more than in most other countries (Federal Committee, 1992c:36; World Bank, 1992a:10; forest data were from an international forest survey commissioned by the United Nations in 1988).

## Laws and Regulations

A substantial but largely ineffectual body of laws, regulations and other administrative policies was enacted during the Communist period to address these problems. Four major environmental laws were passed between 1967 and 1977: the Air Purity Law (1967), Water Act (1973), Agricultural Land Protection Act (1976), and Revision to the Forestry Act (1977). The Air Purity Law established administrative authority for standards, inspection and fines in the Ministry of Forestry and Water Economy, and introduced a system of fines for enterprises that exceeded air pollution standards. The Agricultural Land Act provided compensation to farm organizations for economic losses due to conversion of agricultural land to other uses; the Water Act instituted fees and fines for pollution of both surface and ground water; and the Forestry Law was supposed to protect forest land from construction encroachments as well as industrial pollution (World Bank, 1992:5-6). The principles of responsibility for environmental protection, and even for prevention of environmental deterioration by removing its causes (rather than just treating its consequences), were in fact clearly stated in official state policy documents during the Communist period, and in the Czechoslovak Constitution. Somewhat more detailed description and citation of the laws and regulations of this period may be found in Hrbacek et al. (1989:143-150).

Various technical standards were also promulgated in the 1970s, and by 1984 over 350 regulations had been issued for this purpose, addressing topics ranging from protection of agricultural and natural lands to industrial pollution limits. On paper, many of these standards were actually more stringent than U.S. or West European standards, at least on paper. For example, the 24-hour standard for $SO_2$ was only 150 ug/m$^3$, compared to the U.S. equivalent standard of 365 ug/m$^3$ (World Bank, 1992:9). These laws had some beneficial effects: data show a 50 percent reduction in average dust levels by the mid-1970s, for instance, and a dramatic reduction in dust levels in the three most polluted districts of Prague (State Commission, 1990). A primary cause of the change was a partial shift from coal to natural gas fuels; this appears similar to the conversion that occurred in Pittsburgh (USA) in the 1940s. In 1986 Czechoslovakia signed the Convention on Long-Term Pollution of the Atmosphere, committing itself to a 30 percent reduction in sulfur emissions by 1993 (State Commission for Science, Technology and Investments, 1990:18). Such policy statements and standards conveyed an image to the world that the Communist regimes were at least as committed to environmental protection as the western countries, or even more so.

## Standards versus Reality

In fact, however, most of these paper policies were rarely enforced: the Party-State was both owner and regulator, both polluter and enforcer, both enterprise and banker, and had none of the necessary checks and balances between these functions. In general, these laws had at least three serious deficiencies. First, they left major gaps in authority. The disposal of wastes, for instance, had no legal restrictions whatever, and generally they were simply dumped on industries' land without adequate safeguards against subsequent pollution. Second, some laws actually created perverse incentives, substituting new problems for the old ones. The Air Purity Law, for instance, was often nicknamed the "chimney law," because by regulating only local ambient concentrations it created an incentive to build taller smokestacks, worsening air pollution and forest damage farther downwind (note that similar problems resulted from U.S. air pollution laws of the same period). The Agricultural Land Act, by requiring mitigation for losses of agricultural lands (usually to development or surface mining), created incentives for conversion of agriculturally marginal but ecologically fragile lands to agricultural use (for instance wetlands and forested hillsides). Third, enforcement and penalties for violations were inadequate to nonexistent. Many municipalities and enterprises, for instance, simply obtained "exceptional permission" for continued noncompliance: between 1973 and 1990 approximately 2,500 exemptions were reportedly granted for ongoing water pollution violations, and one may presume similar patterns for air. All exemptions were officially withdrawn in October 1990, but many sources remained out of compliance even though they were now subject to penalties.

## Pollution Charges and Penalty Fees

For at least the past 20 years, Czechoslovakia used emission charges and economic penalties rather than "command and control" regulation or criminal sanctions as its primary public policy tools for pollution control. Enterprises causing pollution were required to pay a basic charge per ton for all emissions of regulated pollutants, and to pay additional penalties for emissions exceeding the standards. This approach has been advocated by western market economists for years, and has now become a centerpiece of environmental policy reform proposals in the U.S. and West European countries (see e.g. Dudek et al., 1992). In Czechoslovakia, however, it failed to prevent or correct some of the worst pollution practices in the industrial world. According to Hrbacek et al. (1989:148), annual reviews by the public prosecutors revealed that the "national committees" which were actually responsible for administering pollution charges and fines rarely punished anyone for failure to prevent or correct pollution, often intervened to have fines reduced or forgiven entirely, and only occasionally used their legal authority either to stop construction or production or to forbid operation of a factory due to health dangers.

There were several reasons for this poor performance. One was the nature of a centrally planned economy: prices, costs and revenues ("profits") were merely accounting entries in an administered system, and so fines and emissions charges, if they were paid at all, were simply built into the enterprises' budgets and transferred to the ministry that charged them without necessarily producing any incentive for reduction of the pollution. Even firms that may have wished to control their emissions, moreover, often could not obtain the necessary equipment on the domestic market and had no access to hard currency with which to import it.

Two other reasons, however, were more generally applicable: first, paying the charges cost less than correcting the emissions; and second, the charges gradually became a decreasingly significant consideration even when they were collected, due to inflation and to the relatively greater increases in the costs of other factors of production over the years after the charges were set (e.g. in 1967 for air pollution).

## Air and Water Protection Funds

A related policy innovation was the creation of special State Funds, which were nominally separate from the general government budget, to pay for air and water protection projects and several other purposes. Revenues for these funds were derived from emission charges and from penalty fees for emissions in excess of environmental protection standards. These revenues amounted to a reported 100-140 million Czech crowns (kcs.) per year during the years 1985-1988 for air protection, out of a total of some 220-238 million kcs. in fees and penalties imposed; water fund revenues were reported at about 1.1 thousand million kcs. per year, including some 35-74 million kcs. in fines (World Bank, 1992b:37, 39, 56-57). The sixth 5-year plan (1976-1980) was the first to include a chapter on environmental protection with quantified targets; subsequent plans allocated increasing amounts of funds for this purpose. For example, seventeen thousand million Czech crowns were allocated by the government from these funds for environmental protection expenditures during 1986-1990. In principle, such funds provided a mechanism for linking financial support for management and restoration directly to charges for the most severe damage. In practice, however, the real uses of these funds were not transparent, the figures cannot be confirmed, and a large fraction of the budgeted funds apparently was never spent. Only 7,000 million kcs. were reported actually spent for this purpose during the 1981-1985 period, for instance, and many of the projects that were initiated were not completed or brought into operation (World Bank, 1992b:10).

*Communism or Industrialism?*

The severity of many of Czechoslovakia's environmental problems was not unique to the formerly Communist countries. As the World Bank commented in a recent report,

In the period between the end of the Second World War and 1970, Western democratic economies showed many of the same problems as those just described. These included: investment in heavy industry; extensive use of coal as an energy source, with concomitant air pollution; careless farming practices such as overuse of fertilizers; dumping of garbage and sewage into surface water bodies; insensitivity to air pollution associated with motor vehicle emissions; and so forth. Indeed, many environmental problems related to these pollution sources remain (World Bank, 1992a:2).

In 1988, for instance, total $SO_2$ emissions for Czechoslovakia were less than those of Great Britain and Spain, and not much higher than Italy's. Its emissions of some other pollutants, such as oxides of nitrogen, were lower than those of France, Germany, Italy, and Great Britain (Federal Committee for Environment, 1992c:18). Extreme values of air pollution from particulates were more serious in Prague than in most European cities, but lower than in either Athens or Madrid; for $SO_2$ they were higher than most but lower than in Milan (Hughes, 1991:Fig.5-7). Its rivers were seriously polluted, but so were the Rhine and other west European waters.

Czechoslovakia ranked sixth in Europe in chemical fertilizer usage, at 338 kg per hectare -- a major cause of ground water contamination -- but the Netherlands ranked first with 346 (World Bank, 1992c:73).

These comparisons must be interpreted cautiously, and are in fact the subject of continuing scientific debate: data estimates may not be of equal accuracy among countries, annual and national averages may have different consequences than intense exposures in smaller but heavily populated areas, and so forth. Given these comparisons, however, one cannot reasonably assume that the overthrow of Communist rule will by itself lead to adequate policies for environmental protection. Communist policies could easily be replaced either by a naive assumption that free markets would solve these problems without government intervention, or simply by such strong political opposition to government authority that environmental damage by businesses would go unchecked in the name of economic progress. However, the current restructuring process also offers a unique opportunity to integrate environmental considerations into the fundamental structure of an emergent market economy and legal structure. The key questions are whether this outcome will occur, and how it can be encouraged.

## Revolution, Disclosure and Documentation

The Communist regime fell in November 1989, in what was called the "velvet revolution" due to its relative nonviolence; six weeks later Vaclav Havel was elected president of the country, and genuinely democratic multi-party elections were held in June 1990 and again in June 1992. As in several of the other countries of central and eastern Europe, opposition to the environmental damage of the Communist regime was one of the important issues around which anti-government opposition could unite (see e.g. Draper, 1993a). Many of the environmental problems were in fact widely known by the public, and even documented in official government records known to many scientists. A key comprehensive report on the environment (the "White Book"), for example, was published in September 1989, before the revolution, jointly authored by the Ecology Section of the Biological Society attached to the Czechoslovak Academy of Sciences, a working committee of the Advisory Council for Economic Research, and the Czech Union of Protectors of Nature. However, these problems had never been fully acknowledged publicly by the government.

An immediate policy step by the new government, therefore, was to end the secrecy and make public officially the full extent of what was known about these problems. The first such document was a report in May 1990 entitled *The Environment in Czechoslovakia*, which was actually a reprint of the "White Book" mentioned above (State Commission for Science, Technology and Investments, 1990b). Later in the same year a similar but more detailed volume was published for the Czech Republic (Czech Republic, Ministry of Environment, 1990), and in 1992 three updated and more detailed national reports were published. The first of these was an updated version of the 1990 report, with additional data (Federal Committee for Environment, 1992b); the second was prepared by a group headed by Bedrich Moldan, who was author of the 1990 report on the environment of the Czech Republic and also Czech Environment Minister (Federal Committee for Environment, 1992c). The third was a detailed atlas of environmental and health conditions in the Czech and Slovak Republics (Federal Committee for the Environment, 1992a).

## Political and Economic Reconstruction

In the wake of the 1989 Revolution, the new leadership of Czechoslovakia attempted to implement six complicated and interdependent agendas at once. First, they *democratized* government, ending the political monopoly of the Communist Party, allowing multiple parties and holding free elections. Second, they sought to create a *market economy* in which prices are set by the forces of supply and demand rather than administrative decisions: most price controls have now been removed. Third, they sought to substitute *private property* ownership for state ownership of most lands and economic enterprises, both by sale and by restitution to pre-Communist owners; this change would also force enterprises to operate on the basis of market prices, hard budgets, and economic efficiency and competitiveness. Fourth, they opened their borders to *freer trade*, by taking steps to make their currency convertible on world markets, and began actively seeking to attract investments by foreign firms as well as export opportunities. Fifth, they sought to *decentralize* governance, reducing the authority of the central government and increasing the responsibility and autonomy of both the two Republics and of municipal governments. Finally, they sought to *protect the environment*, including both cleaning up past pollution hazards, reducing ongoing ones, and preventing new ones.

It is not yet clear, however, what mixtures of these agendas are mutually consistent and effective.

### Freeing of Prices

In shifting to market prices, for instance, it was hoped that the economic reforms would themselves have major benefits for environmental benefits, by forcing closure or fundamental restructuring of the most environmentally damaging heavy industries, which had operated with heavy hidden subsidies in the forms of below-market costs for energy and resources and essentially free rights to pollute the air and water. According to a World Bank prediction, for instance,

The introduction of market forces will cause many obsolescent industries -- often the worst polluters -- to fail. Raised prices will encourage conservation and thus reduce consumption and the environmental consequences of power and mineral production. One estimate is that policies to reduce subsidies in these sectors alone will reduce $SO_2$ emission by at least 25 percent in 1993 (World Bank, 1992a:iii-iv).

In practice, this has not yet occurred. The gross national product did in fact drop by an estimated 16 percent in 1991, and by another 15 percent in 1992, and this was expected to worsen in late 1992 as privatization took hold and the new owners began economic restructuring in earnest (Commission of the European Community, 1992). The recent slowdown in economic production is believed to be reducing emissions at least somewhat. Unfortunately, however, some of the most environmentally harmful industries reduced their production the least -- most remained in the subsidized state sector -- and the consumption of energy resources decreased at a slower rate than did industrial production.

It appears that the initial reform steps, particularly the devaluation of the currency, did not stimulate energy savings but rather stimulated the export of raw materials and metallurgical products to the west, both of which place heavy pressure on the environment, in order to cushion the economic impacts of privatization and freeing of prices. As of September 1992, the president of the central bank reported that gross domestic production was expected to be higher than 1991; and for the first eight months of 1992, Czechoslovakia actually ran a $1 billion *surplus* in its balance of payments, instead of the $600 billion deficit that had been expected, and its inflation rate had been stabilized at less than ten percent (Federal Committee for Environment, 1992b:37).

One reason for this continued reliance on high-pollution heavy industry appears to be that energy prices were still below world market levels, even though they had risen substantially from their previous subsidized levels. A more important reason, however, was the country's deeply-rooted socioeconomic dependence on those industries, a problem which has counterparts in the western capitalist economies as well. Western countries too have found the down-sizing and conversion of government-sponsored military industries an economically painful and politically difficult challenge, and have steadfastly resisted recommendations to eliminate many subsidies which are both economically unjustified and reward environmentally harmful activities over less-damaging alternatives. Examples include below-market pricing of water, timber and minerals; subsidies for road construction, agriculture, and energy extraction; and others. There is every reason to assume that the Czech and Slovak Republics will continue to experience similar political pressures even as market economies, and that these pressures will continue to be compounded by the hardships of the current transition.

An even more fundamental question, however, is the assumption that market-based pricing alone would provide adequate environmental protection under any circumstances. In theory this could be correct if market prices included all environmental costs and benefits, but in practice markets have consistently and notoriously failed to do so. Markets left to themselves create powerful incentives to seek short-term self-interest, at the expense both of unpriced costs to others and of longer-term and cumulative impacts on the society as a whole. Every market economy has found it necessary to use active government intervention -- either regulations, economic incentives, assignment of liability, direct provision of services, or combinations of all of these -- to prevent and correct the many environmental damages and opportunity costs that markets otherwise ignore. Clearly the reduction of energy subsidies is a step in the right direction, but there is no good evidence from the experience of market economies that the freeing of prices alone will clean up or prevent pollution.

*Privatization*

Czechoslovakia moved rapidly with the privatization of small businesses, but the privatization of larger industries has been slower and more limited. By the end of 1991 over 20,000 small businesses had been registered as private enterprises, including many small retail and service shops (note however that many other registrations may have represented merely would-be individual entrepreneurs seeking the tax status and other legal advantages of businesses) (Johnson, 1992). Larger enterprises could be privatized either by direct sale, by auction, by distribution of shares of stock, or by transfer to municipalities or other institutions; by early 1992 nearly 11,000 privatization proposals had been submitted, although many of the latter were competing proposals for the same enterprises (Mejstrik and Burger, 1992). Some key environmentally important enterprises, however, such as some public utilities, were reserved from privatization; and others, such as some major heavy industries, may have had too costly a combination of scale, obsolete capital facilities, and liabilities to attract sufficient interest from investors.

By mid-1992, informed observers estimated that as much as $5 billion of new investment commitments had been made by foreign firms in Czechoslovakia just since the beginning of that year. This process was slowed by uncertainties surrounding the recent restructuring of national governance, but it still represented a significant pattern of ownership change and foreign investment.

Privatization projects are among the truly fundamental decisions that will shape the future environmental conditions in the Czech and Slovak Republics, and they involve at least three key environmental questions. First, what rights do the new owners have to ownership and use of natural resources, and on what terms? Second, what responsibility do they have for cleaning up past pollution on the site? And third, what regulatory expectations will they be required to meet for minimizing current and future pollution, and on what time schedule?

Finance and privatization officials, particularly in the Czech Republic, have taken the position that the economic reform and privatization processes themselves must come first, and that environmental protection for the most part must wait until the economy has been revived. For at least three reasons, however, this policy may be unwise.

First, the environmental deterioration that has been documented represents not merely an absence of amenity, but also real economic and social costs. People whose health is impaired by air pollution or other toxic exposures are sick more often, are less productive workers, and require costlier health care and social services. The economic values of the countries' natural assets, such as forests, fisheries, potable water, arable land, and both natural areas and cultural landmarks valuable for tourism, are reduced or even destroyed by wanton extraction, pollution and waste disposal. A severe air pollution inversion in February 1993, for instance, necessitated cutbacks in power production, restriction of activities by both businesses and individuals, inability of aircraft to land in Prague, and even barring of automobile traffic from downtown Prague (Czech Republic, Ministry of the Environment, 1993). The Czechoslovak Academy of Sciences has estimated that environmental damage cost the country's economy at least 5-7 percent of its annual gross domestic product in the 1980s, and the World Bank has estimated that the full value could be as high as 10-13 percent (World Bank, 1991:111), though other observers have offered lower estimates. Industries with serious environmental liabilities are less attractive to investors than they would be otherwise, and contaminated sites and water sources cannot be used for economic purposes for which they are otherwise suited. The economic development potential of whole towns or regions may be seriously hampered by the burden of past and continuing environmental deterioration, and by the necessity for more costly infrastructure investments (such as water purification technology, or procurement from distant sources) to remedy or adapt to it.

Second, many of the economic reform decisions will themselves fundamentally determine the environmental results for the next generation or longer. If private investors purchase a state-owned chemical factory, for example, their proposal implies important assumptions about how much liability for cleaning up past pollution they expect to bear, and how much the government -- which may not have the resources to clean it up -- will assume. It also involves assumptions about what air and water pollution standards their new capital investments must meet, and how soon; what changes in land uses they will be permitted to make; and what their costs of energy, water, waste disposal, and other natural resource inputs are likely to be. For example, one firm which recently invested in a Czech company continued to discharge significant amounts of phosphorus to the nearby river; if the government had negotiated seriously on the subject, the firm probably could have been persuaded to install more effective wastewater treatment facilities, but it reportedly was never asked to do so (personal communication, J. Scherer, 2/93). The government's real environmental policies will thus be predetermined in important ways by its negotiation of these privatization agreements.

Third, a policy of deferring environmental protection considerations may even be self-defeating in terms of its own objective of attracting rapid capital investment from abroad. Major investors are experienced enough to expect that higher environmental standards will be instituted within the economic life of today's capital investments, and they also are well aware that it is far more expensive to retrofit environmental controls on plants once constructed than to design efficient processes from the outset. What they need therefore is not weak controls in the short term, but clear policy signals about the longer term, such as negotiated but binding timetables for cleanup to specified standards, and explicit policies that air and water quality standards will rise at $x$ rate until at the end of $y$ years they are identical to those of the European Community. They are likely to delay major capital investments until these expectations are made clear.

One promising environmental policy initiative was enactment of a law requiring environmental audits ("assessments of environmental obligations") as an element of all privatization proposals, including those in which foreign firms become partners in investment.[5] This requirement provided an important potential policy tool for integrating environmental considerations into the privatization process, both by compelling disclosure of environmental issues and by providing a point of leverage for those interested in or responsible for such issues. In practice, however, the Environment Ministry has neither the authority nor the staff resources to monitor these assessments systematically; and the law itself provided no formal role for public review or challenge of them. Informed observers therefore have expressed skepticism about its effectiveness.

Notwithstanding this assessment, some positive initiatives have also occurred, such as the emergence of a new business-oriented not-for-profit organization, the Czech Environmental Management Center. Created in 1992, this center is funded by a combination of government support and business dues; by 1993 it had over 50 member companies, about half either private or in process of privatization, and was engaged both in efforts to improve environmental policies and in promoting cost-effective pollution prevention, safety engineering, and environmental management in industrial operations.[6]

An important unanswered question about the privatization process, however, concerns the true economic and environmental objectives of foreign partners in such agreements. In some cases they are seeking normal participation in already-viable, attractive, and perhaps undervalued businesses, or businesses that could easily be made viable through better management and access to external markets. In other cases, however, they are buying into enterprises widely characterized as technologically out-of-date, overstaffed, poorly managed, and environmentally damaging. What is it then that they are buying? For some it may be an inexpensive but technically skilled labor force; for others it may be an inexpensive site; for still others, a stake in emerging free markets in central and eastern Europe.

At least three other possibilities, however, deserve careful public scrutiny. One is that they are buying pollution havens, or at least short-term profit advantages of less stringent environmental controls. That is, they may be buying opportunities to continue using technologies and processes that are no longer environmentally acceptable in the U.S. or European Community, at least for a few more years that the Czech and Slovak governments might be willing to concede them in order to attract their investments.

Second, they may be buying control over potential competitors. Informed observers have commented, for example, on purchases of Czech and Slovak enterprises by similar European firms, with clauses requiring that not only their products but also their production processes must meet all EC standards before their products could be sold in western markets -- thereby preempting any price advantages that those enterprises might obtain from less stringent standards.

A third possibility is that they are buying in to control the traditionally monopolized markets of central and eastern European countries themselves. One western firm, for instance, reportedly bought a Slovak pulp and paper industry that was actually quite technologically modern, closed it and substituted production from its competing Austrian plant in the Slovak firm's market, costing Slovakia the loss of both jobs and profits. Any of these possibilities raise serious questions of both environmental and broader economic, social and political consequences.

## Decentralization

Interacting and overlapping with privatization is the process of both reducing and decentralizing the functions of government. Decentralization of governance requires not merely a dismantling of centralized authoritarian structures, but also the *creation* of effective governments at more decentralized levels. Many environmental problems arise from localized sources -- such as municipal and home heating, air pollution from small and medium-sized businesses, discharge of untreated wastewater, and dumping of waste materials -- and require similarly local solutions. Local and regional governments must therefore be created that are accountable to their citizens rather than merely to central government; that have stable and appropriate sources of revenues, and authority to tax and borrow in order to fulfill their responsibilities, rather than merely budget allocations from the central government; and that have the competence, information and expertise, as well as the authority, to manage the responsibilities, services, and environmental hazards of their jurisdictions.

Under the Communist regime, municipal governments in Czechoslovakia developed in a very weak form. They were local administrative agents of the national Interior Ministry, rather than autonomously financed; they were politically accountable to a structure of "National Committees," which functioned as administrative organs of the Communist Party and its ideological positions and internal politics, rather than being accountable to their citizens directly; and they were dominated in practice by the power of the local state industries, which reported directly to more powerful national ministries. A further nuance of these patterns was that the staff bureaucrats were often far more influential than the elected officials. The strength of these patterns varied somewhat by area, by the type of industries present, and by the size of the municipality (Dostal et al., 1992:17-32).

Beginning in 1989 several important changes occurred, though their full consequences for environmental and other policies are not yet clear. First, the National Committees were abolished, along with the regional level of administration (*kraje*) as a whole; district-level National Committees were transformed into state administration offices, and may themselves be abolished in the future; and municipal governments became locally elected, relatively self-governing units. In effect, the Republics now have only two levels of governance, the Republic and its municipalities; and the municipalities themselves range in scale from Prague to the smallest hamlet.

Second, important authorities over environmental and other policy sectors began to be decentralized to municipal governments. Municipalities were authorized to regulate small sources of air pollution, for example, and to manage smog emergencies. They were also authorized to manage municipal property; to allocate their shares of environmental emissions fees and penalty revenues, as well as service charges for water, energy and waste disposal; to approve private enterprises' waste management plans, and to prepare their own; to manage land uses and urban services within their jurisdictions; and to provide a variety of environmental services, such as wastewater and waste management and central heating. Other services, such as water supply (previously administered by regional river basin authorities), were opened to privatization proposals, which might be submitted either by municipalities or by entrepreneurs.

By 1992, many changes in elected officials and political processes had begun to result from democratization, but key changes in resource allocation and administrative capabilities are still incomplete. First, the municipalities are now independent in principle, but still dependent on top-down budgeting for most of their resources, and these resources have been severely reduced due to national revenue shrinkage; and they remain severely circumscribed in the sources and rates of taxation they may use. Most of the municipalities therefore have not yet developed a sense of true autonomy and self-reliance. Second, there is still considerable uncertainty about what authorities and property ownership rights the local governments really have, and how creatively they may use them. Third, municipal elected officials are still new both to their offices and to electoral processes and political relationships, while many staff officials are still in place in the administration; it will take time to establish new patterns of governance.

An enduring challenge is that both the small size of many municipalities and the limited revenue sources available to all of them carry inherent limitations on the efficiency with which public services can be provided. As of 1991 Czechoslovakia had over 8,600 municipalities, less than 400 of which had more than 10,000 residents, and 75 percent of which had populations of less than 1,000 (Dostal et al., 1992:28). Communities of this size have continuing needs for regional service systems and for more centrally provided technical assistance. Even larger cities do not yet have access to commensurately larger sources of revenues: municipal bond markets do not yet exist, for instance, and sources of revenues are still contingent on both unresolved authority to set taxes and fees and the uncertain financial futures of newly-privatized local businesses (let alone those still in state ownership).

There have been reports of growing numbers of policy initiatives by some of the larger towns and cities. At least 15 cities, for example, were exploring replacement of municipal technical service functions (municipal cleaning and household waste management, for instance) by private contractors (Price Waterhouse, 1992). Others had at least commissioned studies of key environmental problems such as air and water pollution and waste management, and of options for solving them. Some have taken more aggressive initiatives, such as restricting sulfur content in fuel deliveries, investments and incentives for conversion from coal to less damaging fuels, and others. The pace of such initiatives may increase, at least in the larger municipalities, with the expected enactment of the tax law in 1993.

Four important issues remain unresolved, however. The first is how much autonomy the municipalities will have to finance their services, and to levy taxes and charges to pay for capital as well as current expenditures. The second is how much authority (and resources, and political will) they will have to control land development patterns, and particularly to protect important environmental values of undeveloped lands.  The third is how to provide necessary environmental services to small communities, and adequate technical competence for environmental management even in larger ones; and the fourth is how to solve environmental problems that require regional-scale rather than merely municipal solutions.  At least the latter two problems will probably require development of intergovernmental agreements, new regional service districts, or other governance institutions which are more accountable to local voters than in the past but can also pool financial resources from larger areas and provide economies of scale in facilities, markets, and management expertise.

*Democratization*

Environmental concern was a powerful and central issue for the Civic Forum movement that toppled the Communist regime, and remained an important issue during the euphoria of the first years of independence.  It was popular both within these countries and abroad to characterize the formerly Communist countries as the most polluted places on the face of the earth, and to blame all these conditions on the evils of authoritarian Communist governance and centralized economic planning.  This message was a powerful force for mobilization of the public against the Communist regimes, and it was actively publicized by the new governments and widely disseminated by western journalists.  Polls at the beginning of 1990 identified environmental issues as the most important concerns of 83 percent of the population; and respected scientists who had courageously exposed these issues before the revolution were named to appointments in the Environment Ministries and other government agencies (State Commission for Science, Technology and Investment, 1990a:20ff).

Since then, however, public attention and the political agenda have been dominated by the pocketbook issues of economic restructuring and by the combination of ethnic, ideological and economic issues associated with Slovak separatism.   By 1992 much of this movement had substantially dissipated as a political force, both in Czechoslovakia and even more so in other East European countries, and almost all of the strong environmental leaders were no longer in government.  One reason was clearly that the political agenda had become dominated by economic fears and by jockeying for power among new political factions (including the use of strict anti-Communist criteria to discredit even many of the leaders of the reform movement, who appeared "tainted" by circumstances before the 1989 revolution).

A related problem is that citizen environmental organizations have not yet developed strong patterns of political power and influence: many exist, but most operate in the mode of nature education rather than policy advocacy, leaving elective politics dominated by the economic and political transformation itself rather than its environmental implications. Realistically speaking, U.S.-style non-governmental policy advocacy groups simply are not yet an effective presence in Czech or Slovak politics: they do not yet have either the resources, the experience, or the political or legal leverage to exert effective influence on public policy decisions. Like the United States under President Carter, moreover, key environmental leaders after the velvet revolution moved inside the government (even though lacking in power and sometimes in political effectiveness relative to the economic ministries) -- leaving their organizations both needing to develop new leadership, and reluctant to challenge their friends in government as strongly as they did the Communists.

A more general problem for environmental politics was that many of those most severely impacted by pollution lacked strong political influence and remained fearful and divided. A large fraction of the population in both Republics, including many of those most at risk from environmental health damage, continued to depend for employment on the old state-supported heavy industries whose fate was still unresolved. In North Bohemia, for instance, a large proportion of the population was relatively poor in both income and education, heavily dependent on employment in mining and heavy industry, and fearful of losses both of jobs and even of small subsidies they had received in the past for living in an extremely polluted region.

In the 1992 elections, the centrist Civic Forum movement lost power dramatically to the more clearly defined ideological positions of both the free-market oriented party of Vaclav Klaus in the Czech Republic (Federal Finance Minister and now Prime Minister) and the Slovak nationalist Movement for a Democratic Slovakia of Vladimir Meciar in the Slovak Republic (Draper, 1993b). In general, therefore, environmental protection has been clearly overshadowed for the time being by economic and governance issues and by the conservative market-oriented policies of the Czech government.

## Environmental Protection Laws and Regulations

In the spirit of environmental concern that was an integral element of its "velvet revolution," Czechoslovakia acted immediately to begin establishing new administrative organizations, laws, regulations and other policies for environmental protection. These actions included initiatives at the Federal, Republic, and municipal levels of government, and amounted to more specific reforms in environmental law than in any other country in the region (Bowman and Hunter, 1992).

It is not yet clear, however, to what extent these policies will be more effectively implemented and enforced than were their predecessors under the Communist regime, or to what extent they will simply continue the previous disparities between nominal and actual policy under the new banner of a free-market economy. Nor is it clear whether they will adequately prevent or remedy the new environmental threats that may be created by a free-market economy itself.

## Federal Policies

Both environmental and other public policies in Czechoslovakia were shaped by its federal structure of governance. Laws enacted by the Federal Assembly provided a general framework of principles, which must be implemented by more detailed bodies of legislation in the Czech and Slovak Republics. A small Federal Committee for Environment was established in 1990, chaired by a Federal Minister of the Environment (Josef Vavrousek); but each Republic had its own environmental ministry (in Slovakia, the Slovak Commission for Environment), and most of the authority for more specific legislation, as well as the staff for implementation and enforcement, lay with the Republics.

One important general policy was the Charter of Fundamental Rights and Freedoms, enacted 9 January 1991 to be incorporated into the Federal and Republic constitutions.[7] Similar in concept to the U.S. Bill of Rights, the Charter granted all citizens basic political rights, including access to information, participation in the administration of public affairs, and public health. Its environmental section asserted rights to live in a favorable environment and to obtain information about environmental conditions, and prohibited all persons from damaging the environment, the diversity of species, or cultural monuments beyond statutory limits. It further asserted that these rights could be enforced by citizens through an independent Constitutional Court. The effectiveness of these provisions in practice, however, has not yet been tested (Bowman and Hunter, 1992).

Between 1989 and 1992 both the Federal government and its constituent Republics enacted a series of policies, laws and regulations dealing specifically with environmental protection. The main goals and principles were laid out in two documents in early 1990, the Czechoslovak national report to the international conference in Bergen on issues related to sustainable economic development, and shortly thereafter the *Concept of State Ecological Policy* (State Commission, 1990a, 1990b).

Both these documents were strongly influenced by the United Nations' Brundtland Commission report *Our Common Future*, which advocated the goal of sustainable development.   They reaffirmed overall government responsibility for maintaining environmental quality in the context of the emerging market economy and more autonomous local governance, with the goals of protecting human health, conserving nature's wealth, protecting cultural and economic values from unfavorable impacts, and protecting the life-giving systems of the planet.   Their basic principles included integrating environmental considerations and incentives into the decisions of all sectors of government and society, not simply setting up a separate ministry for environmental regulation.   They also differentiated two main tasks:  cleaning up existing pollution problems, and preventing new ones by closing cycles of production and consumption.

By mid-1990 a joint statement of ecological programs and projects was issued by the Federal and two Republic ministers, defining their mutual priorities, and by later 1990 the three governments requested that the World Bank, European Community and United States work with them in a joint environmental study.   A draft was completed in January 1991, and a final version   in   January   1992 (World   Bank,   1992a:viii-ix).     This   study recommended   highest   priority   for   reducing   air   pollution   and   food contamination, followed by water and solid waste issues; primary emphasis on "hot spots" of particularly serious contamination, such as North Bohemia and   large   cities;   and   careful   balance   among   scientific,   recreational   and commercial   objectives   in   policies   for   nature   conservation   and   protected natural areas.   In reality this statement was never a true joint program, but rather a general statement of good will and shared ideals, without serious programmatic elements, resources, or effective priorities.   It was nonetheless a symbolic step forward.

Beginning   in   1991   the   Federal   Assembly   enacted   a   series   of   new environmental laws, the general goal of which was to bring Czechoslovakia's legal system as much as possible into harmony with the emerging system of the European Community (EC).   Among the most important new laws were those establishing principles for managing wastes, air, and the environment in general; others proposed included laws on water management, land planning and development, forest management, farmland preservation, mining, and animal rights (Federal Committee for Environment, 1992b).   A proposal for an Environmental Code of Ethics for businesses was also circulated for discussion, modelled on a "Business Charter for Sustainable Development" already adopted by the Executive Board of the International Chamber of Commerce and endorsed by 50-100 corporations and other organizations (Federal Committee for Environment, undated).

The Waste Act, passed in May and effective in August 1991, was Czechoslovakia's first law on wastes.   It established the responsibility of generators for the safe management and disposal of their wastes, and banned all waste   imports,   exports   or   transit   without   specific   approval   of   the responsible government agencies. All waste management projects and facilities must also be approved; generators must also have approved waste management plans, waste treatment facility operators must have approval for all collection, purchase and treatment of wastes, and all must keep records. Violators were subject to steep penalty fees depending on the degree of potential hazard involved.[8]

The Clean Air Act, passed in July and effective in October 1991, significantly updated the 1967 air purification law. Unlike the 1967 law, it established a general obligation for air quality protection by mobile in addition to fixed sources; it required best available technology for all new sources, and required large and medium-sized sources to cease operation during severe air pollution conditions. Siting and construction of new sources, as well as changes in technologies or fuels for any large or medium-sized sources, required approval by the air protection authority of the appropriate Republic; those authorities could impose fines or if necessary order restriction or cessation of operation. Special restrictions were also authorized for areas and time periods subject to severe smog conditions.[9] In October the Federal Committee issued a regulation implementing this law, by listing regulated pollutants, categorizing their sources, and setting Federal emission limits; these limits were specific to each major industrial technology, and were modelled on those of the European Community.[10]

Finally, the Environmental Law, effective in December 1991, established a systematic set of general principles for environmental protection based on the concept of sustainable development. These principles included a general public right to obtain information about environmental conditions, causes, proposals to change them, and agency responsibilities. They also established individual liability both for compliance with environmental standards and for preventing or remedying environmental damage. The law required environmental impact assessments for all projects, plans, and other proposed economic activities that required government permits; it authorized financial penalties for any violations that caused ecological damage, and cessation of activities causing serious peril or damage; and it authorized the use of taxes and other economic incentives to achieve environmental protection and natural resource conservation.[11]

In 1992 the Federal Assembly incorporated an unspecified "ecological tax" into drafts of its proposed Tax Act. The initial proposal was for a tax on hydrocarbon fuels with offsetting reductions as rewards for energy efficiency improvements; as of February 1993, however, these proposals had never been further developed and had not been incorporated into the Tax Act revision.

Czechoslovakia also received substantial assistance on environmental issues from a number of other nations and international organizations, although it was more reticent to accept loans for this purpose than were some of its neighbors. The Joint Environmental Study (mentioned above) provided a primary framework for this assistance; grants included over $45 million from the Poland-Hungary Assistance for Restructuring the Economy (PHARE) program of the European Community, $15 million from the U.S. Agency for International Development, and bilateral activities with more than half a dozen of its West European neighbors.    There was some disgruntlement among knowledgeable Czechs, however, that too much of this assistance was paid to support consulting firms from the donor countries rather than to finance actual solutions or even to build capabilities for such analyses within the Czech and Slovak Republics themselves.    Two major World Bank loans were also being negotiated, one for rehabilitation of the North Bohemian power plants ($246 million) and the other for implementation of the Joint Environmental Study Action Plan ($150 million). As of early 1993, however, these agreements had not yet been finalized, reportedly because of reluctance by the government to increase its foreign debt by guaranteeing them:  the government appeared to prefer loans only for commercial purposes with short payback potential (Evans, 1992).   One additional initiative was a $4.3 million grant from the Global Environment Facility to plan and manage five national parks that had particular significance for ecological conservation.

     Taken together, these laws and initiatives laid out the beginnings of a set of principles for environmental protection in Czechoslovakia.    While differences in detail will undoubtedly emerge between the two now-separate Republics, these laws provided a substantial common core of principles underlying the environmental policy systems of both.

## Republic Policies

Following enactment of each of the Federal laws in 1991, the two Republics began introducing their own more detailed statutes and regulations.   The Czech Environment Ministry, for instance, almost immediately published its *Rainbow Programme -- Environmental Recovery Programme for the Czech Republic*, an idealistic agenda for integrating environmental objectives into the larger process of economic and political reform.   Two basic principles of this programme were that "the polluter pays," through a system of economic charges and penalties roughly proportionate to the severity of the damage, and that these revenues would then be earmarked for environmental cleanup projects.   A third major element would be the development of regional programs for integrated management of pollution problems in severely impacted areas, such as North Bohemia and Ostrava.   The immediate priorities of the ministry were to get basic environmental protection laws enacted, establish the Environment Fund and other economic instruments, and set up the basic state administrative apparatus, information systems, and initial projects to get the agenda underway.   More detailed agendas were also proposed for each major environmental issue and program element (Czech Republic, Ministry of Environment, 1991:7-14).

In the case of waste management, for instance, the Czech Republic passed two additional statutes and an official public notice between July 1991 and January 1992, establishing the agencies responsible for waste management, the responsibilities of municipalities and district councils (as well as private generators) to prepare waste management plans and programs, and the topics that must be covered in such plans. Plans were specifically required to include methods for preventing and minimizing waste generation as well as for safe disposal. The laws also established the levels of charges for waste disposal and of penalty fee levels for violations: charges were to be set by the Republic but were to accrue to the community in which the waste disposal site was located, while penalty fees were to be divided between the Czech Republic, the district and the affected community depending on which initiated the action. As yet, however, no sites had been officially approved for hazardous waste disposal, and no comprehensive list of cleanup priorities for existing sites had been completed.

An important and related issue was the liability of new owners for past pollution by the enterprises they were purchasing. This issue was important both for environmental reasons and as an obstacle to investment. There were three possible solutions: either the new owners would be liable, in which case they would want to buy at steeply discounted prices; or the government would agree to be liable, though investors might not trust such agreements since the government lacked resources to fulfill such commitments and might change its mind once they invested; or some third-party insurance fund could be made responsible, supported by a portion of the revenues from privatization. In each case, however, a key issue was how much cleanup would be accepted as sufficient: this issue has stymied the United States even after investment of far greater sums of money, and was an obvious problem for Czechoslovakia as well.

In 1992 the Czech government passed a resolution on this issue, declaring that the government would assume full responsibility for any off-site environmental damages and that liability for on-site damages would be transferred to the purchaser, provided that 50 percent of the purchase price would be put in escrow for remediation of these on-site damages. The escrow would reimburse the purchaser for costs of government-ordered on-site cleanups within one year, up to a maximum of 50 percent of the purchase price. This solution left major issues unresolved, including off-site problems, problems that might cost more than 50 percent of the purchase price to remedy, and problems that might not be evident until more than a year later; but it did appear to provide some legal reassurance to purchasers to smooth the process of privatization. Before it was implemented, however, the resolution was repealed and replaced by a revised policy under which purchasers remained responsible for all costs of cleanup unless specific exceptions were approved by the government.

In the case of air quality, overall policy was made the responsibility of the Republic itself, as were regulation of large and medium-sized sources of air pollution in the Czech Republic and emission charges for large sources.[12] Medium-sized sources and their emission charges, as well as the quality of fuel supplies, may also be regulated by district councils; small sources, as well as smog emergencies and mobile source inspections, were the responsibility of municipal councils.[13] Emission charges were to be phased in over five years (later amended to seven years), rising from 30 percent of the prescribed rates in 1992 to 100 percent in 1997; the rates were to be differentiated between common pollutants (particulates, $SO_2$, $NO_x$, CO and hydrocarbons) and three classes of hazardous pollutants to which much higher rates were applied. Fifty-percent surcharges were also specified for emissions exceeding environmental standards (e.g. 1,000 to 3,000 kcs. per ton for common pollutants, versus 10,000 kcs./ton for arsenic, lead and phenols and 20,000 kcs./ton for cadmium, mercury and benzene). A cautionary note is in order due to the frequent absence of monitoring equipment, which limits the verifiability of emissions and thus the enforceability of accurate charges.

Enforcement for major air pollution sources in the Czech Republic was assigned to the Czech Environmental Inspection, a semi-autonomous professional unit attached to the Environment Ministry, and for medium-sized sources to the district councils; for small sources it was assigned to the municipalities. The Slovak Republic took a somewhat different position: noting that many of the small sources were operated by the municipalities themselves, particularly municipal heating plants, they assigned enforcement even for local sources to a separate staff of professionals who were to be based at the local level but reported directly to (and were paid directly by) the Slovak Environment Ministry.

The Czech Environment Ministry was also authorized to define areas requiring special air protection measures and to specify what those measures should be. In December 1991 it issued a decree establishing an initial set of such regions: these included national parks and protected landscape areas, spas, urban zones in which pollution was restricted, and perhaps most important, a series of cities and districts already seriously impacted by air pollution (for example Prague, Ostrava and Plzen, most of the districts of North Bohemia, and others).[14] Aside from smog emergency response actions, however, the nature of these special measures had not yet been spelled out. Additional legislation might be necessary to authorize the sort of regional air quality management district that could impose ongoing planning controls and incentives in a seriously impacted area such as North Bohemia.

In October 1991 the Czech Republic created a Republic Environmental Fund, supported by revenues from fines and penalties for exceeding air and water quality standards, and consolidating the several funds that had previously existed for environmental purposes (air quality, water quality, groundwater extraction, and conversion of agricultural lands). The levels of these penalties were expected to rise from initial levels of about 30 percent of estimated compliance cost, to 100 percent by 1997; and similar fines for waste disposal violations were also to be collected beginning in 1992. Note again, however, that underreporting of emissions as well as pleas of economic hardship and corresponding changes in statutory timetables could easily erode these revenue expectations. In addition, pollution generators reportedly can negotiate discounts from these charges to finance investments in pollution control (though it is not clear that they could do so to pay for more general modernization measures that would reduce or prevent pollution rather than treat it). Of these amounts, 60 percent was to be awarded for projects in the districts from which it came; total revenues to the fund in 1992 were expected to be about 2.2 thousand million kcs. ($70-80 million) (personal communication, J. Scherer, 2/92).

With these revenues, the Fund could make grants, loan guarantees and actual loans for environmental projects, with low interest for projects using the most environmentally beneficial technology. All awards must be for actual project implementation, not studies; the Fund may pay up to 50 percent of all projects approved, and up to 90 percent for projects of non-profit organizations. Projects must be approved by both the local government and the staff of the Environment Ministry, and among these, allocations were to be made by the Environment Minister on the recommendation of a board of fifteen members, two-thirds from industry and banking and the rest representing key ministries.

Finally, in April 1992 the Czech Republic passed a law requiring environmental impact assessments for a wide range of construction activities as well as of policy "concepts" and of selected products, and disclosure of these assessments for public discussion and comment (Czech Republic, National Council, 1992).

In the Slovak Republic the agenda of policy development was broadly similar, but its content differed in some specifics. The Czech Republic assigned high priority to air pollution and industrial wastes; the Slovak government gave greater initial emphasis to water pollution. Slovak environmental responsibilities initially were assigned to the Slovak Commission for Environment, and only in 1992 was this Commission redefined as a ministry. Both Republics enacted basic statutes for major environmental problems, and requiring such activities as environmental audits of businesses; Slovak legislation initially moved somewhat slower than in the Czech Republic due to the division of legislative matters into a greater diversity of separate statutes.

## Unresolved Issues

By the beginning of 1993, the Czech and Slovak Republics were continuing to move through a process of political and economic transformation which had the potential, though not a guarantee, of positive environmental results as well. Democratization of governance had occurred, and privatization was proceeding, albeit slowly; investment commitments had begun to flow into the countries; and systems of environmental laws and regulations and agencies were being put in place. But all these processes were still in their early stages; unemployment was expected to worsen, at least temporarily, as enterprises were restructured for greater profitability by their new owners; the separation of the Czech and Slovak Republics slowed this process, and might lead to new differences in policy directions between the two Republics; most actual correction of environmental damage and its causes had not yet occurred; and the influence of both environmental concerns in general and the Environment Ministries specifically had visibly declined, even in the face of highly visible and continuing problems.[15] Important issues remained to be resolved.

### Privatization and Modernization

A first key unresolved issue is how successful the economic transformation itself will be in modernizing the technologies that have caused much of Czechoslovakia's environmental pollution, from large-scale industrial and agricultural processes to local power plants, apartment buildings, and individual homes and appliances. Vaclav Klaus, Czech prime minister and formerly Federal finance minister, used as one of his campaign slogans the phrase "Economy for Ecology:" in his opinion, reconstructing the economy was itself the key to generating the wealth and the technological modernization necessary to clean up and protect the environment. While economic revitalization is surely a necessary condition for environmental protection, however, it may not by itself be a sufficient one: much depends on *what kind* of economic reconstruction the actual pattern of public policy choices encourage. One pattern would provide strong incentives from the start for environmentally sustainable technologies and businesses, and actively phase out environmentally damaging processes. An alternative pattern, however, would ease the shock of reconstruction by keeping environmentally damaging industries going with marginal improvements and delayed compliance. The real policies may well be a case-by-case mix of these two patterns, depending on the economic stakes and opportunities, but the environmental results may be anything from gradual progress to continuation of serious pollution.

## The Energy Future

Nowhere are these issues more important than in the energy sector, where a large fraction of the country's most serious pollution problems continue to be caused by a combination of large coal-fired power plants, coal-based municipal and household heating, inefficient use of energy in industry as well as housing and appliances, and rising use of motor vehicle transport. Some municipalities are actively promoting local fuel conversion from coal to gas, but this process is costly and also dependent on supplies (all currently from Russia) and prices. At worst, many of the smaller sources, which are collectively the most serious sources of urban air pollution, may simply continue to pollute. Most of the large power plants also continue to operate: will they be closed, and alternative sources of energy developed? Or will the government undertake the enormous cost of retrofitting them with air pollution control technology? And what can be done to seriously reduce energy demand, in ways consistent with development of an environmentally sustainable economy?

In January 1993 Prime Minister Klaus announced his policy that the controversial new Temelin nuclear power plant should be completed. This plant, nearly completed to a traditional Soviet design, is to be upgraded to western safety standards, and has been promoted by its supporters as a replacement for current large-scale combustion of brown coal. To have this effect, however, will require additional policy commitments that have not yet been made, to phase out the coal-burning plants and also to convert from coal to electricity or other cleaner fuels at the household level. The full costs of the nuclear fuel cycle, and of the environmental and socioeconomic remediation of the coal economy, also have not yet been systematically addressed.

## Implementation of Environmental Protection Laws

Perhaps no question is as crucially important, yet so far uncertain, as whether the Czech and Slovak Republics will in fact effectively implement and enforce the environmental standards, fees, and penalties which they have now established on paper. Both have now created environmental ministries, and have enacted a new set of environmental protection laws, regulations, charges, and penalties. It is worth repeating, however, that the Communist regime also had stringent environmental standards: the problem was not a lack of standards, but insufficient penalties, and exemptions rather than enforcement. Will the new standards be enforced, or will new exemptions be granted in the pursuit of foreign investment, or compliance deferred in the terms of privatization agreements? Will the new charges be high enough to be a serious incentive for correction of pollution -- for instance, will they be indexed to inflation? Or will they still be too low and serve only as a revenue source -- or even be reduced or waived if a firm pleads that it can not afford to pay them and will go out of business? In a true market economy, changes in the cost of any factor of production may cause marginal firms to fail, and such failures are in fact routine: to the extent that environmental damage causes real harm and costs, companies that cannot operate economically while paying for them should be allowed to fail, not propped up by hidden subsidies in the form of exemptions.

The long-term answers to these questions are not yet clear, but the evidence through early 1993 is not encouraging. Air pollution compliance deadlines have already been extended by two years, and business complaints were increasing concerning the economic burden of emissions fees on their profits; political pressure could conceivably lead to future easing of penalties or rollbacks of deadlines, especially as the deadlines loom closer. The environmental ministries and inspectorates were understaffed to begin with, and have become even more so due to budget cuts and staff reductions: they do not appear to have adequate staff even to sustain systematic monitoring and inspection programs, let alone to review the many privatization proposals in which key environmental consequences are in fact being determined. Since the elections of 1992, responsibility for the Czech Environment Ministry was given to the Christian Democratic Party, a coalition partner of the ruling free-market oriented party of Klaus; party appointees have now replaced most of the previous senior staff, and in the view of most observers the influence and effectiveness of the ministry have greatly declined. Following a severe air pollution episode in February 1993, Prime Minister Klaus visited the ministry for a full morning meeting, but used the occasion mainly to claim political support for the Temelin nuclear power plant, to criticize the ministry for addressing economic questions rather than merely ecological ones, and to suggest that its staff be reduced from 250 to 80. Some environmentally important policy-making responsibilities, such as regional planning, have been transferred out of the environment ministry to ministries charged with promoting economic development. Many of the most capable individuals, meanwhile, especially those who speak English, are being lured away from government by more attractive opportunities in the rapidly-growing private business sector.

*Tax Policies and the Environment*

A major policy element still being implemented is the new set of laws on taxation. These were supposed to include explicit provisions for ecological and environmental taxes, but in fact such proposals have not yet been developed and incorporated into legislation. Equally important will be what incentives the tax system as a whole might create towards either positive or negative environmental impacts, and how much authority the municipalities will have to levy their own taxes for environmental or other purposes. At the municipal level, the capability of municipal governments to deal with their own problems will depend heavily on how much discretion they are given to issue bonds and seek loans, and to use taxes, service charges, and other revenue sources to repay them on a self-financing basis. Like several other key issues, these policies remain very much a moving target.

## Preservation Opportunities

The environmental protection agenda between 1989 and 1992 was heavily dominated by the "hot spots" of extreme pollution.    These are without question important and immediate risks for health and ecological damage, and they were therefore designated as official priorities by agreement among the governments of Czechoslovakia, the two Republics, and major international agencies.

Equally important, however, is the unique but brief window of opportunity that now exists to protect valuable natural areas and environmental resources before they are opened to private economic claims.   The North Bohemian highlands, for example, despite severe damage in some areas due to air pollution, are still a region of exceptional beauty and distinctive character, especially the famed "Bohemian-Saxon Switzerland" of the Ore Mountains near the German border.  South Bohemia too has a distinctive beauty that is so far largely untouched by industrial or commercial exploitation; and the High Tatra range of the Carpathian Mountains in Slovakia ranks among the beautiful and relatively undeveloped mountain areas of Europe.  A hotly debated set of preservation opportunities also exists in the past military zones, particularly along the borders (the former "Iron Curtain"), which had been closed to the public and therefore remained virtually untouched for four decades or more.

All these areas have been preserved so far in part by the absence of private market opportunities to profit from them, and in some cases by Cold War military exclusion policies.  Proposals currently under discussion, however, would designate small areas for greater protection as national parks, but at the same time remove the "landscape-protected" status from larger regions, opening them to commercial privatization without any new system of land use controls yet in operation to protect important environmental features and values.

In every market economy, the protection of environmental values of land and water resources has required action by governments, both to preserve especially important areas outright and to set and enforce reasonable restrictions on how otherwise private lands may be used. These purposes can be accomplished far more economically before privatization rights are transferred.  Most of the magnificent U.S. national park and national forest systems, for example, were made possible only by the fact that the land was already in government ownership:  repurchase would have been far too costly to protect more than a small fraction of them, and some are still marred by pre-existing resource extraction rights that the government cannot afford to buy back.

The same principle is true of smaller areas, such as urban green space, wetlands, floodplains, and steep slopes; of water resources, which have not only commercial but also ecological and aesthetic values; and of other lands that have important ecological functions, such as watershed protection and wildlife habitat.  Such lands do not all have to be kept in government ownership outright.  However, it would be far less costly and more effective to establish clear easements or other appropriate restrictions on their future use as conditions of initial privatization transactions, than to have to fight endless political battles to re-regulate them later after initial terms of ownership and economic expectations have been established.

Two important and urgent issues remain unresolved, therefore.  The first is to assure systematic preservation of the most important lands and waters throughout the Czech and Slovak Republics, while that could be accomplished by policies reserving them from privatization.  The second is to institute an explicit system of land use controls to protect essential public values of other lands and waters even after they are privatized, and to establish clear responsibilities as well as rights for those who will become private owners of lands and their environmental resources.  Examples might include easements or similar use restrictions for such areas as wetlands, wildlife habitats and biocorridors, and especially beautiful landscapes.

## Markets and Government

Market forces by themselves should not be expected to solve all environmental problems, and may cause new ones.  Free markets in the United States and West Europe have caused many of the same environmental problems that centralized planning caused in Czechoslovakia, including air and water pollution, waste dumping, and destruction of natural landscapes for industrial and commercial purposes.  The reason is that while some environmental damages are caused by inefficiencies that businesses themselves gain from correcting, many others can be prevented only by additional capital investments and operating costs.  If businesses are not required to pay the full costs of environmental damage, therefore, market processes themselves create powerful incentives to cause such damage, since businesses that do not pay to prevent such damage will have lower costs and thus be able to undersell those that do.

Free markets will also bring new environmental challenges of other kinds. Rising economic standards may bring more efficient use of resources, but it may also bring increases in the total use of resources, with more waste materials to manage. More automobiles will increase pollution in cities and towns, and also increase pressure to build more highways which will change the landscape in important ways. Even tourism, which is often considered a "clean" industry, brings environmental pressures and impacts that must be carefully managed.

A critical new problem is the catastrophic decline of government revenues and budgets, at all levels, for essential public-sector environmental services: public transportation, water supply and wastewater management, urban maintenance, monitoring and enforcement of standards, health services, planning and management of ecologically important lands, and others. Some of these services were underfunded to begin with, but even others that were previously better managed, such as a reasonably comprehensive nationwide public transportation system, are now seriously deteriorating.

Environmental protection in a market economy therefore requires active roles for government as well:   at least to set clear standards for environmental conditions that must be maintained, to make sure that prices reflect the full costs of environmental resources and damages (such as energy and water use and land destruction by mining), and to provide essential environmental services that markets normally do not (such as wastewater and waste management and management of ecologically important lands).

The creation of new models for effective government, however, is not yet as far advanced in the Czech and Slovak Republics as the freeing of prices and the privatization of businesses. Budgets and staff have been severely cut back in many areas, particularly those that do not have immediate market benefits but are for that reason important government services even in market economies:   examples include environmental protection, education, and scientific research. Municipal governments' budget allocations from the Republics also have been cut back severely, but without yet having the capability or the authority to raise their own revenues to comparable levels. More generally, privatization risks abdicating important government responsibilities that will then have to be recreated or bought back from the private sector at higher cost later. It is important that the Czech and Slovak governments develop as quickly as possible a clear understanding of the essential interdependence of free markets and effective governance, and act to redevelop competent government capabilities as equal partners and complements to their emerging market institutions.

*Politics of Environmental Protection*

A major unresolved issue is the extent to which widespread public support for environmental protection can be mobilized and sustained as an effective political force. This will be important for Republic-level issues, such as implementation of environmental laws and economic reconstruction of energy and other industries, but it will also be essential for local as well as many crucial regional-scale issues, such as cleanup of the Elbe (Labe) river, restoration of the mined areas of North Bohemia and industrially impacted areas of both Republics, minimizing environmental impacts of new transportation infrastructures, and others.

Environmental groups in West Europe and the U.S. provide important but imperfect models for this role. They have developed strong capabilities for critical analysis and political advocacy, but they have often been more effective in opposing environmentally damaging proposals than in building broad coalitions to support better ones. The current transformation offers opportunities to try to build such coalitions, but they will require a strong combination of creative and well-reasoned ideas and consensus-building across traditional adversary groups in order to realize them. These conditions do not yet appear to exist.

# Conclusion

Several tentative conclusions may be drawn from this overview. First, the environmental problems of the formerly Communist countries are themselves more complicated and heterogeneous than they have generally been portrayed in the western news media. Serious pollution is not a uniform and constant condition, but one that exists in particular regions and "hot spots" and in particular seasonal conditions -- making it more difficult to sustain nationwide political pressure for cleanup, especially in stressful economic conditions. The pollution that is most hazardous to health, moreover, is not always the same as the pollution damaging forests downwind, raising dilemmas for resource allocation: is it more important to spend scarce finances on cleaning up local heating sources in cities or big power plants in the country (or how can the resources be found to do both?) Finally, many areas are relatively unpolluted or even pristine, yet the danger of losing these assets to rapid and uncontrolled commercial exploitation has so far received little attention compared to privatization and pollution.

Second, the actual effectiveness of the new governments' environmental policies cannot yet be determined. Basic environmental statutes and regulations now exist on paper, but they are not yet widely implemented and enforced; and the congruence between these regulations and the companies' restructuring and investment plans, let alone the governments' commitment to compliance on behalf of the many heavy industries that are still state-owned, cannot yet be estimated. Both governments and the emergent private sector are still sorting out their intentions.

It is possible therefore that these policies will create a climate of expectations that will lead to good-faith implementation and compliance. It is also possible, however, that the gap between paper policies and actual practices which existed under the Communists may simply continue, or that implementation will be limited to collection of relatively painless pollution charges as a revenue source rather than enforcement of serious cleanup of emissions.[16] In the mean time, most environmental improvements to date are probably attributable more to economic slowdown and plant closures rather than to incentives or investments in more environmentally efficient production; and most of the state-owned, high-pollution heavy industries continue so far to operate as before.

Third, the fact that Czechoslovakia's severe environmental problems arose under Communism does not mean that free markets and democratic governance will necessarily correct them. Among the formerly Communist countries, the Czech Republic appears to be one of the most aggressively committed to promoting a rapid transition to free-market institutions, and to counting on the economic dynamism of that transition to provide the financial resources for subsequent environmental cleanup. These policies will probably reduce those environmental problems that result from inefficient subsidies and obsolete technologies, but it may continue or even exacerbate those that result from firms' pursuit of their own short-term self-interest. In the rush to privatize as rapidly as possible, moreover, important forms of environmental protection that *did* exist under the former regimes may be lost to the commercial pressures of a market economy. This would be unfortunate. Bowman and Hunter (1992:976-979) also note more general concerns about the mixed motives and conflicts of interest on the part of many western investors, governments, and lending institutions.

Finally, even the most optimistic results of the initiatives so far underway do not yet reflect any coherent government policy of pursuing environmentally sustainable development. Some of its initiatives are arguably consistent with such a result, but so far only incidentally: the primary policies of both the current governments reflect older and more limited approaches to economic reconstruction. This may in fact be one of the most fundamental opportunities that is being overlooked in the current government approaches to the post-Communist transition.

A series of important questions therefore deserve increased attention concerning the nature of the transition and its environmental consequences. For instance:

· Pollution under the Communists was not always caused by a lack of laws and standards, but often by their non-enforcement. What timetables for environmental cleanup and compliance are being incorporated into the industrial privatization agreements now being signed, and into resource allocation decisions for the major facilities that remain in government ownership; and how will they be enforced? Or will market-driven economic conditions merely replace Communist production quotas as a new rationalization for continued pollution?

· "Market-oriented incentives" for environmental protection are being promoted as an alternative to "command-and-control regulation." Which of these would in fact promote ecologically sustainable patterns of development, and which of them are appropriate for adoption in the Czech and Slovak Republics, under what conditions, and at what levels?

· Privatization, of land and public services as well as of industrial facilities, is being advocated as a remedy for both the economic and the environmental failures of Communist management. What safeguards will prevent the competitive destruction of environmental amenities that occurred in market economies before the introduction of government restraints?

· Decentralization of governance is both delegating and abdicating many environmental protection functions to the municipalities. Which of these do they have the authority, the capability and the resources to carry out, and what additional initiatives will be necessary to deal with larger regional-scale needs and with problems too costly or technically complex for municipalities to solve?

· Finally, environmental concern was an important element of the political mobilization that overthrew the Communist regime, and this led many western advisors to seek to create ongoing western-style environmental advocacy groups in Czechoslovakia. What powers, resources and other conditions are necessary for such groups both to wield effective influence and to achieve beneficial results? And what other initiatives will be necessary for the emergence of durable political constituencies for environmental protection and sustainable development in the Czech and Slovak Republics?

The answers to these questions are not yet clear, but the questions themselves represent an important agenda for further study and debate, both in the Czech and Slovak Republics and in the other transitional economies of central and eastern Europe.

## Acknowledgements

The research presented in this chapter was derived from a project on Economic Approaches to Environmental Policy in the Czech and Slovak Republics, sponsored by the U.S. Agency for International Development, whose support is gratefully acknowledged. Thanks are also due to Bedrich Moldan, Vladimir Balaban, Lubomir Paroha, Petr Sauer, Margaret Bowman, John Dernbach, Richard Liroff, James Scherer, and others for their review of drafts and helpful comments and corrections.    All opinions and any remaining errors are solely the responsibility of the author.

## Notes

1.  In the Czech Republic, for example, all the Federal laws were reaffirmed in a special act passed at the end of 1992.

2.  A Czech economist notes that even Russian experts at the time advised against intensive development of brown coal-fired power production, but were overruled by the political authorities (personal communication, P. Halouzka, 12/92).

3.  Hrbacek et al. (1989:139, 149), for instance,  noted that while the first program for construction of sewage plants was initiated in 1958, the cost of constructing wastewater facilities was not an obligatory part of the budget approval process for new housing facilities, but was allocated separately from an "ecological investments" category which had low priority.

4.  This problem is obvious in the crumbling facades of many historical buildings; it is less obvious but equally serious in such statistics as an estimated 30 percent rate of water losses from leaking pipes (Bedrich Moldan, personal communication 2/93).

5.  Federal Assembly, Amendment to the Privatization Act, Law No. 92/1991 dated 18 February 1992, adding section 6(a); implemented in the Czech Republic by Methodical Instruction dated 18 May 1992, and in the Slovak Republic by a similar instruction dated July 1992; see also Federal Committee for Environment, 1992b:38.

6.  For instance, the Center sponsored a detailed study of the impacts of emissions fees on half a dozen major firms, and found that at each of several levels their actual effects on net profits of the firms were minimal.  One policy proposal from the Center, therefore, was to index

emissions fees to inflation; another was to establish by law a set of consensual principles for evaluating and integrating environmental laws and policies.

7.  Federal Assembly, Constitution Law 23/1991, January 1991; cited in Federal Committee for Environment, 1992b:119.

8.  Federal Assembly, Waste Act of 22 May 1991.

9.  Federal Assembly, Clean Air Act, Law No. 309/91 dated 9 July 1991.

10. Federal Committee, Regulation of 1 October 1991 Concerning Law No. 309/1991 dated 9 July 1991.

11. Federal Assembly, Environmental Law, Law No. 17/1992 Coll., dated 5 December 1991.

12. Note an important source of possible confusion, both in understanding and in policy approaches: in Czechoslovakia "large" and "medium-sized" sources were legally defined by their power output, not by the actual magnitudes of their emissions.

13. Czech National Council, Act on the State Administration of Air Protection, No. 389/91 dated 10 September 1991.

14. Decree No. 41/1992, dated 23 December 1991.

15. A vivid confirmation of the continuing reality and cost of environmental contamination was a severe week-long air quality inversion which occurred in Prague and North Bohemia during the first week of February 1993. The most serious reported inversion in at least three years, this condition caused peak 30-minute $SO_2$ levels as high as 2,588 ug/m3 near Usti nad Labem (North Bohemia) and 1,851 in downtown Prague. For comparison, maximum concentrations allowed under national and World Health Organization standards for 24-hour concentrations are 150 ug/m3 (Ministry of Environment of the Czech Republic, Material Pro Poradu Vedeni MZP CR -- Informace o prubehu smogove situace v 6. tydnu roku 1993, od 1.2. do 8.2.1993, Cj.:520/498/93).

16. One important further step, for instance, would be to index emission charges and penalty fees to inflation, so that they retain their intended incentive effect rather than becoming less and less effective over time.

# References

Bowman, Margaret and David Hunter. 1992. Environmental Reforms in Post-Communist Central Europe: From High Hopes to Hard Reality. *Michigan Journal of International Law* 13:921-980.

Carter, F. W. 1984. "Pollution in Prague: Environmental Control in a Centrally Planned Socialist Country." *Cities*, February 1984 (Butterworth & Co., Ltd.).

Commission of the European Community. 1992. *The European Economy*, Supplement A.

Czech Republic. Resolution No. 287, dated November 2, 1990, Concerning the Environment-Depolluting Program of the North Bohemian Region for 1991-1995 and with the Prospect until the Year 2000.

Czech Republic, Ministry of the Environment. 1990. *Environment of the Czech Republic*. Parts I-III (*"Blue Book"*).

Czech Republic, Ministry of the Environment. 1991. Decree No. 41/1992, dated December 23, 1991, establishing regions requiring special air protection, and establishing principles for the creation and operation of smog regulation systems and other measures for protection of the air.

Czech Republic, Ministry of the Environment. 1991. Public Notice No. 401/1991, dated August 16, 1991, on waste management programs.

Czech Republic, Ministry of the Environment. 1991. *Rainbow Programme: Environmental Recovery Programme for the Czech Republic.*

Czech Republic, Ministry of the Environment. 1993. *Informace o prubehu smogove situace v tydnu roku* 1992, od 1.2. do 8.2.1993. Cj.:520/498/93.

Czech Republic, National Council. 1974. Law on the State Administration of Water Management, Law No. 130/1974 Sb.

Czech Republic, National Council. 1991. Act on the State Administration of Air Protection and Charges for the Pollution of Air, Law No. 389/91, dated September 10, 1991.

Czech Republic, National Council. 1991. Act on the State Administration of Waste Management, Law No. 311/1991, dated July 8, 1991.

Czech Republic, National Council. 1992. Act on Charges for the Deposit of Waste, Law No. 62/1992, dated January 22, 1992.

Czech Republic, National Council. 1992. Act on Environmental Impact Assessment, Law No. 244, dated April 15, 1992.

*Czechoslovakia: Official Economic Magazine of Czechoslovakia*. Prague, February 1992.

Dostal, Petr, Michal Illner, Jan Kara and Max Barlow (eds.). 1992. *Changing Territorial Administration in Czechoslovakia: International Viewpoints*. Amsterdam: Institute for Social Geography, University of Amsterdam.

Draper, Theodore. 1993a. A New History of the Velvet Revolution. *New York Review of Books* 40/2:14 ff. (January).

Draper, Theodore. 1993b. Goodbye to Czechoslovakia. *New York Review of Books* 40/3:20-26 (January).

Dudek, Daniel J. et al. 1992. Environmental Policy for Eastern Europe: Technology-Based Versus Market-Based Approaches. *Columbia Journal of Environmental Law* 17/1:1-52.

Evans, Sheila S. 1992. *Environmental Activities in the Czech and Slovak Federal Republic*, 1991. New York: Charter 77 Foundation.

Federal Assembly. 1973. Water Law No. 138/1973, dated October 31, 1973.

Federal Assembly. 1991. Clean Air Act, Law No. 309/91 dated July 9, 1991.

Federal Assembly. 1991. Environmental Act, dated December 5, 1991.

Federal Assembly. 1991. Waste Act of May 22, 1991.

Federal Assembly. 1992. Amendment to the Privatization Act, Law No. 92/1991, dated February 18, 1992.

Federal Committee for Environment. (n.d.). Draft Environmental Code of Ethics: Principles for Environmentally Sustainable Management For All Business in Czechoslovakia, Domestic or Foreign.

Federal Committee for Environment. 1991. Regulation on Air Pollution Control, Implementing Law No. 309/91, dated October 1, 1991.

Federal Committee for Environment. 1992a. *Atlas of the Environment and Health of the Population of the Czech and Slovak Federal Republic*. Brno and Prague: Geographical Institute of the Czechoslovak Academy of Sciences.

Federal Committee for Environment. 1992b. *National Report of the Czech and Slovak Federative Republic to the United Nations Conference on Environment and Development*.

Federal Committee for Environment. 1992c. *State of the Environment in Czechoslovakia*.

Federal Committee for Environment, Ministry of Environment of the Czech Republic, and Slovak Commission for Environment. 1990a. *Concept of State Ecological Policy*.

Federal Committee for Environment, Ministry of Environment of the Czech Republic, and Slovak Commission for Environment. 1990b. *Ecological Programmes and Projects, Czech and Slovak Federative Republic*. July 1990.

Federal Committee for Environment, Ministry of Environment of the Czech Republic, and Slovak Commission for Environment. 1991. *State Programme of Care for Environment/Structure of Programme*. April 1991.

Hrbacek, Jaroslav; Binek, Bedrich; and Vaclav Mejstrik. 1989. Czechoslovakia. Chapter 9 in Edward J. Kormondy (ed.), *International Handbook of Pollution Control* (New York: Greenwood Press), pp. 137-151.

Hughes, Gordon. 1991. *Are the Costs of Cleaning Up Eastern Europe Exaggerated? Economic Reform and the Environment.* Washington, DC: World Bank.

Johnson, Simon. 1992. Private Business in Eastern Europe. In K. Froot and J. Sachs (eds.), *The Transition in Eastern Europe.* Washington, DC: National Bureau of Economic Research.

Mejstrik, Michal and James Burger. 1992. Privatization in Practice: Czechoslovakia's Experience from 1989 to Mid-1992. Chapel Hill, North Carolina: Kenan Institute of Private Enterprise.

Moldan, Bedrich and Jerald Schnoor. 1992. Czechoslovakia: Examining a Critically Ill Environment. *Environmental Science & Technology* 26(1): 14-21.

Myjavec, Josef and Anna Szulenyiova. 1992. Economic Instruments for Environmental Protection in CSFR. In *Economy and Environment.* Bratislava: Ustredny Ustav Narodohospodarskeho Vyskumu pobocka Bratislava (UUNV).

Price Waterhouse. 1992. *Privatization of Solid Waste Management Services in Czechoslovakia.* Draft Final Report to USAID/Czechoslovakia, dated April 10, 1992.

State Commission for Science, Technology and Investments. Department of the Environment. 1990a. *Czechoslovak National Report on Issues Relating to Sustainable Economic Development.* Paper presented at the Conference on Action for a Common Future, Bergen, Norway, May 8-16, 1990.

State Commission for Science, Technology and Investments. Department of the Environment. 1990b. *The Environment in Czechoslovakia.* (Josef Vavrousek and colleagues, authors).

World Bank. 1991. *Czechoslovakia: Transition to a Market Economy.* Washington, DC.

World Bank. 1992. *Czech and Slovak Federal Republic Joint Environmental Study.* Report No. 9623-CS, Vols. I and II. Washington, DC.

Chapter 3

# ENVIRONMENTAL AFFAIRS AND THE EMERGENCE OF PLURALISM IN POLAND: A CASE OF POLITICAL SYMBIOSIS

Stanley J. Kabala

In Poland, intensive industrial development and severe ecological deterioration came together to produce environmental activism and force the restructuring of an unbalanced economy. The environmental conditions behind this phenomenon affect a significant segment of the country and place a substantial share of its people and land in danger. Twenty seven regions, comprising 11.3 percent of the country's land area and 35.5 percent of its population, had been officially identified as "areas of ecological hazard." Five of these regions -- Gdansk Bay on the Baltic, the Legnica-Glogow copper district in western Poland, and the contiguous Krakow-Katowice-Rybnik coal and steel belt in the south -- experience degradation and pollution so acute as to be classed as "areas of ecological catastrophe" (Kassenberg and Rolewicz, 1985). Losses resulting from the degradation of the environment and waste of natural resources have been estimated at over ten percent of the country's annual national income. Official sources in the late 1980s came to refer to the "ecological barrier to the development of the country" (Kassenberg, 1990; Kostrowicki, 1988; Symonowicz, 1985).

To the extent that Poland's environmental problems can be traced to the prediliction of its socialist system and planners toward heavy industry as the motor of development, those problems can be diagnosed as economically-induced. That is, they are not the result of such factors as climatic disturbances or demographic pressure on the country's ecological balance. Rather, the way in which the country chose to conduct its agriculture, its industry, its energy policy -- in sum, its pattern of development -- is the principal material cause of the environmental deterioration it is experiencing. Yet, as Polish analysts over the decade of the 1980s insisted with progressively decreasing obliqueness, the country's environmental crisis was not to be alleviated merely by making adjustments to the country's economic structure. It could be corrected *only* by reconstruction of the political system so thorough-going as to result in a new system, not a reformed one. This is the political root of the country's environmental problem which we will examine in this chapter.

The broader field in which this theme played itself out was the country's political history since the institution of communist rule after World War II. Poland's post-war pattern of development failed to satisfy both the economic needs and the social aspirations of its people. After forty years of

*A. Vari and P. Tamas (eds.), Environment and Democratic Transition, 49–66.*
© *1993 Kluwer Academic Publishers. Printed in the Netherlands.*

unsuccessful effort on the part of the regime to reconcile communist ideological objectives with a highly resistant society, unresolved fundamental economic and political problems generated a degree of social discontent sufficient to force radical transformation of the country's political system.

Environmental issues formed a subset of the range of issues that constituted the agenda of the political opposition in Poland in the 1980s. In this period, the issue of environmental pollution acted, in effect, as a delegitimizing agent in Polish society, seen by the Polish people as an indictment of the policies of the regime in all quarters. The result was that society perceived the remedy for environmental degradation to be not specific governmental regulatory policy, but rather fundamental transformation of the country's political system. The result was that the country's state socialist system began to diverge from the pattern of monolithic organization characteristic of communist social systems. In this state of change, areas of discourse and decision making are removed from the control of the communist regime and placed in a new realm of public, pluralist political activity.

## Environmental Policy Before 1981

The issue of the environment passed through four periods in post-war Polish history. The first two decades after the Second World War were not an environmental era at all anywhere in the world, but rather were characterized variously as a time of nation building (in the Third World), economic reconstruction (in war-damaged countries), social reconstruction (in communist states), and pursuit of economic growth (universally). In Poland, as throughout most of the world during the quarter century following the Second World War, economic growth dominated the imaginations of countries and governments, with environment rarely given a second thought. The idea that natural processes possessed carrying capacities that could be exceeded to the detriment of both nature and man was unknown or not regarded as pertinent.

The beginning of the 1970s marked the beginning of environmental concern in Poland, but this concern was largely limited to a small segment of the Polish scientific community, and general social awareness and activism on environmental issues had yet to appear. There were several reasons for this. First was the limited supply of information on environmental issues. The government's tight control of technical data prevented its dissemination to both the public at large and a significant portion of the scientific community. Second was the very limited amount of ecological research being conducted on a subject which was clearly not a high priority for the regime. Third was simply the relative newness of the issue of ecological degradation in a country that had begun to industrialize only thirty years before and where people still equated economic progress with the imagery of heavy industry.

As a result, public concern and activism on the issue had not emerged. The only public organization with an environmental agenda was the official League for the Protection of Nature (Liga Ochrony Przyrody, or LOP), Poland's oldest nature organization. Founded in 1928, the LOP had become, under the country's communist regime, a "transmission belt" for official positions on ecological matters. Its functions were limited to largely traditional activities in nature education and nature conservation and it was not considered by Poles to be a body from which to expect aggressive action (Sadowski, 1981). This situation was in contrast to that in Hungary, where by the late 1970s a ecological awareness was evident in society. The impetus to environmental protection that came from a committed scientific community had no parallel in Poland. One Western observer commented at the time that government environmental programs in Poland, unlike those in Hungary, appeared to be "principally a matter of rhetoric rather than action" (Kormondy, 1980).

In this context, the government's assertions that pollution was the price that Poland had to pay for a rising standard of living received their share of credence. In the 1970s, concern over the environmental costs of industrial development seemed out of place in a time of bustling investment, rising wages, and full shops. The "New Investment Policy" of the 1970s, fueled largely by foreign credits, still boded to bring prosperity to the country and increased legitimacy to its communist government. At the time, environmental concerns were largely minimized by official claims of the need for economic growth and assertions that growth itself would provide the resources for environmental protection. The environmental effects of Poland's decades-long drive to industrialize were ignored or referred to as the price that the nation had to pay for progress. This assertion was to be challenged by the country's environmental movement in the 1980s.

By the last years of the 1980s, the country's communist regime was desperately trying to come to terms with the interlinked problems of environmental protection and economic reform in its final efforts to achieve domestic political stabilization. The emergence of Solidarity in 1980-1981 opened the third era of environmental affairs in Poland, a period in which discussion of environmental questions became an acceptable part of public political discourse, environmental issues made their way onto the national agenda, and environmentalism became a political force. At the same time, in the years up to 1985, political and social activity on environmental issues was accompanied by singularly unconstructive action on environmental protection.

The beginnings of change were marked in 1985 by an official watershed, the creation of a Ministry of Environmental Protection and Natural Resources that gave cabinet status to this issue. The last half of the 1980s was a time when environmentalism asserted itself on several fronts: social awareness, political action, and governmental policy. Events at mid-decade led this observer to speculate that a convergence of circumstances might lead to greater care being given to environmental quality in Poland. Elements identified at the time were the growing recognition by other central European governments of the need for an international response to industrially induced environmental decay; the apparent acceptance by the Polish scientific community of a public interest role in the ecological discussion; the continued existence, activity, and acceptability of public environmental groups; and a government that, although still committed to industrial growth, was open to advice on environmental matters (Kabala, 1985).

The 1984 Munich Conference on Transboundary Air Pollution and the (then) uncharacteristically open participation of Eastern Bloc countries was an example of the sort of external force for environmental improvement to which the Warsaw government might acquiesce. Poland did not join "the 30 percent club" by agreeing to reduce national emissions of sulfur dioxide in 1990 to 30 percent of 1980 levels, claiming economic inability to do so, but its actions on all other international fronts were positive.

The Polish scientists who responded to the need for their technical abilities during the Solidarity era showed no sign of abandoning this style of involvement afterward, especially insofar as it formed a part of their regular professional scientific work. The question was whether the Polish scientific community would play a role in environmental affairs like that played by its counterpart in Hungary (Kabala, 1985).

Activist organizations became a fixture in the environmental arena in Poland even though the extent of their influence was not clear. The Polish Ecological Club spread across the country and constituted a network of active local action groups composed of citizens, scientists, and local municipal officials whose voice was regularly heard on matters of local and national environmental policy. By 1987, the government felt itself obliged to create an official environmental organization, the Social Movement for Ecology, the Ekologiczy Ruch Spoleczny or ERS. The very establishment of the ERS -- with rather prestigious connections to the government and the Polish Academy of Sciences -- is an indication of how important the government found it to appear to be moving positively on ecological matters. (Kabala, 1990) Relative to regimes in other socialist countries of Eastern Europe, the Polish regime came rather late to the idea of using environmental protection as a tool of political legitimation. As we will see, when it tried to do so, the issue played into to the hands of the opposition, for reasons distinctive to Polish politics.

The significance of Solidarity in the history of environmental affairs in Poland is paramount.  Events during the sixteen month era of Solidarity moved in matters of ecology as rapidly as in all other areas of Polish life, and an environmental movement was born that outlived its outlawed parent union.  By the beginning of the 1980s, failed communist social reconstruction and intensively pursued but faltering economic growth culminated in economic collapse, political dissension, and ecological crisis.   In 1980, Solidarity, Eastern Europe's first independent self-governing trade union was born out of these problems.  Solidarity opened up Poland's civic life and released a flow of information on every aspect of Polish society, economy, politics, and recent history.  Virtually no area of affairs was safe from contributions by Poles willing to contribute knowledge in their fields to the general flow.

The environment was no exception.  Subsequent to the government's lifting of its ban on the subject in 1980, heretofore restricted information on the ecological condition of the country flowed into public discussion of environmental matters in popular and scientific fora.  Throughout the country, official data were made public and industrial records made available by administrators, officials, and scientists sympathetic to the aims of Solidarity.   Local chapters of Solidarity investigated and compiled information on environmental abuses.  Number sixteen of Solidarity's 1981 theses included environmental protection and demanded the overhaul of environmental regulations, the reform of industrial accounting to identify environmental losses, the creation of environmental protection funds to be managed by local government bodies, the conversion of polluting industries, and the construction of adequate waste treatment systems.   Despite its eventual suppression, Solidarity had made possible a time of social and political openness in Poland that martial law could not eradicate.   It left behind a Polish ecological movement as well.

A striking illustration of the nature of the times was the publication in mid-1981 of a special edition of the LOP journal, *Przyroda Polska (Polish Nature)* entitled "The State of the Environment in Poland and the Threat to Human Health."   The report compiled the best information existing in Poland on the extent of the ecological threat to the country and its people, presented an estimate of the toll in economic and social welfare that was being exacted, and proposed ways out of the crisis.  As the product of an official, government-sponsored organization, the report forcefully brought to the attention of the Polish people the dimensions of the natural disaster they were facing that the government had seen fit to keep from them.  The data in the special edition went on to be incorporated into other scientific and official reports and found its way into the country's large and effective system of "underground" publications.

The case of the Skawina aluminum plant is illustrative of the impact of the times on environmental affairs. The thirty year old plant located fourteen kilometers to the southwest of Krakow accounted for half of Poland's domestically produced aluminum (Pudlis, 1981). It also generated over 2,000 tons of fluoride pollutant, the extremely toxic residue of the smelting process. Reports prepared in the late 1970s showed that the plant posed a serious health hazard to its workers and to people and farms in an area of some 230 square kilometers surrounding it.

In his study of the Solidarity era, *The Polish Revolution*, Timothy Garton Ash observed the degree to which the Skawina case indicated how Poland's public life during this time resembled that of Western liberal states. In late 1980, a coalition of Solidarity, scientists, local environmentalists in Krakow, and the press demanded that the aluminum plant be closed down. The issue became the major organizing tool of the Malopolska (Krakow) branch of the newly formed Polish Ecological Club (Polski Klub Ekologiczny, or PKE). Krakow's principal daily, *Gazeta Krakowska*, ungagged and operating under a new editor, took up the environmentalist cause against the Skawina works. Continued pressure by Krakow city authorities, the press, and the PKE caused the State Ministry of Metallurgy to completely shut down the plant by January 1981 (Ash, 1982). Despite the reassertion of communist rule in Poland in the mid-1980s the plant was never reopened.

The Skawina episode was the high water mark of environmental activism in communist Poland. Despite the regime's reestablishment of its political authority, that achievement was not contradicted. When martial law was imposed and Solidarity suppressed in December 1981, the fate of its environmental offspring, the PKE, was not clear. Officially registered under Polish law in 1981, in fact the PKE was tolerated by the Polish government and grew into a nationwide network of Club chapters active on regional and national environmental questions (Fura, 1985).

*The Factors of Change*

Three concepts provide a frame of reference with which to examine the sources of environmental deterioration, the process of making environmental policy, and the emergence of environmental activism in Poland: the Stalinist extensive model of development, the behavior of bureaucratic organizations in a field of competitive uncertainty, and the emergence of political pluralism. Let us note these briefly here.

The developmental root of Poland's environmental crisis is the Stalinist model of resource mobilization, the so-called extensive pattern of development undertaken in the Soviet-dominated countries of Eastern Europe after World War II (Fallenbuchl, 1965). This Soviet-style ideological base resulted not only in the intended implementation of the desired extensive pattern of development but also in the creation of other, unintended residual effects, particularly resource inefficiency and concomitant environmental deterioration. The accompanying introduction of Stalinist political and administrative forms into Eastern Europe established a closed political system dominated by an ideologically-motivated strategic elite -- the respective national communist parties. Over time the administrative culture that developed influenced decision-making processes to the detriment of environmental quality even as the harsher aspects of Stalinism disappeared from the Polish (and East European) political scheme (Bojarski, 1986; Drewnowski, 1982). The assured -- but in the Polish case not unchallenged -- decision-making power of this elite in society framed the field on which economic and environmental issues were decided (Misztal and Misztal, 1984).

Within the decision-making structure of communist regimes there was interest group competition. Unlike pluralist political systems, the field of competition was not the public arena but rather the bureaucracy. The ineffectiveness of environmental policy in Poland and its fellow East European states is accounted for by unbalanced competition among interests *within* the state bureaucratic setting in which environmental interests are steadily overwhelmed by industrial interests. In these terms, change in environmental politics in Poland can be seen to have resulted when several things took place. Environmental interest groups emerged in the Polish polity and placed pressure on the regime to initiate change. Environmental agencies in the bureaucracy cultivated constituency power deriving from growing environmental awareness in society at large, which they could tap to counteract the economic power of the ministerial industrial lobby. The effect over time was progressive movement of the locus of interest group activity and competition from *within* the governmental/administrative apparatus *outward* into the emerging arena of public political participation.

This process of change can be viewed as the development of pluralism in a formally non-pluralist political system. Morawska (1988) has applied this concept to discern patterns of decision-making in communist Poland. The slow break-down of Poland's quasi-totalitarian system made way for the competing conceptions and functions that constituted the core behavior of pluralism as an alternative pattern of social organization. "Systemic functional pluralism," according to Morawska, is characterized by tolerance of independent political, economic, and social associations; institutional provision in the polity for their activities and for the reconciliation of conflicting interests; and the existence of channels for voluntary group participation in the policy process. Conflict is resolved to the extent possible through compromise in a context of democratic institutions (Morawska, 1988; Staniszkis, 1984). Ziegler offered a useful description of the relationship between pluralist political forms and environmental affairs in his analysis of environmental policy in the Soviet Union. He pointed out that in authentic pluralist systems, responsibility for the environment is diffused throughout society, and citizens, private firms, and government agencies all may be held responsible for damaging or protecting it. The range and number of independent interest groups concerned with environmental protection in pluralist systems constitutes a broad source of information and political power on environmental questions. In Soviet-type corporatist systems responsibility for the environment, control of information, and the generation of new ideas flowed through the much more narrow channels of the party-state structure (Ziegler, 1987).

The conditions of pluralism, formally nonexistent in communist systems, emerged in Poland during the Solidarity era of 1980-1981 and continued to develop despite regime wishes to the contrary. After Solidarity, through the 1980s the situation remained that of unbalanced competition among interests within the state structure of legislation and allocation. In this bureaucratic arena, economic actors, i.e., the heavy industry lobby, consistently overwhelmed environmental interests, resulting in minimal and ineffective action on environmental concerns. The degree of influence on the government exercised by the economic ministries considerably exceeded that exercised by the Environment Ministry, and budgetary and investment decisions on economic issues continued to be skewed toward industrial activity. The ability of environmental regulatory bodies to constrain deleterious activity by economic ministries was largely neutralized by their political power in the bureaucratic decision-making arena.

The situation changed as a new set of actors entered the policy making arena, progressively developing and using the political tools common to interest groups in pluralist societies. Social pressure for environmental protection derived from the activities of scientists, political activists, the media, and assorted citizens groups, as well as a concerned and dissatisfied public. The government found itself forced to respond to public demands for environmental protection efforts -- a demand in its own right and, in the context of Eastern Europe, merely one of a series of political demands put forth by the opposition. As concern for environmental quality took shape as a new factor in the political equation, the regime reacted by making policy decisions that affected both the environmental regulatory agencies (positively) and the economic ministries (negatively). Discussions of investment planning increasingly took environmental protection into account -- at least on paper. As its constituency grew, the ability of the regulatory agency to directly affect the activities of offending enterprises increased as well. The model that emerged resembled that of political behavior in pluralist systems, where issues are placed on the political agenda not merely by the ruling elite or party and on the government agenda not merely by bureaucratic actors in a closed system, but rather by diverse interest groups acting in the arena of society at large.

An apt backdrop against which to view structural trends in Soviet-type economic and political systems has been provided by Drewnowski (1982) in his characterization of the degradation of basic "economic tissue." Drewnowski asserted that degradation of economic tissue was a condition brought about by key characteristics of the systems themselves and posed fundamental, long-term difficulties for them. He based his assessment on the observation that the social environment exhibits certain basic characteristics. These are that the economy operates within a definite legal system, contracts are enforced, honesty is a prevailing practice, and managers, professionals, and manual workers are generally able to perform their work to the best of their ability. In particular, managers can collect and evaluate information, assess the productive capabilities of their operations and the impact of external factors, understand the directives they receive, and exercise their own judgment (Drewnowski, 1982).

Drewnowski pointed out that economic tissue can be degraded so as to affect the very functions that an economy is supposed to perform, and asserted that aspects of tissue degradation appear to have set in centrally planned systems by the end of the 1970s. "Suppression of truth" is an element of degradation that derives from the belief that Marxist doctrine is right and will prevail over alternative doctrines. The concomitant prediliction to secrecy in Soviet-type systems leads to a reluctance to impart valid information that highly impairs the quality of decision-making that must go on in any case.

Simultaneously, the "eradication of dissent" limits the range of meaningful discussion and destroys the very capacity for self-correction to which socialism has for so long ascribed its theoretical superiority over market systems. The necessary on-going assessment of any activity implies the criticism of what has been done and not done before -- a difficult task in an environment that sees criticism as hostile. The comprehensiveness and centralization of decision-making that was long considered the hallmark of planned economies requires equally comprehensive monitoring and equally available information. Resistance to the flow of information counters and obviates the attribute of centralized socialism that has been its bulwark since Lange: the more nearly perfect system of information and coherent direction that it has at its disposal that is lacking in capitalism (Wright, 1971; Lange, 1938).

Last among the causes of tissue deterioration is the "repudiation of fairness," in society at large and in economic enterprises in particular. Foremost in the institutionalization of unfairness is the demoralizing practice of "nomenklatura," the politically-charged selection of candidates for high positions throughout society on the basis not of expertise but political steadfastness. Administrative damage is engendered by this procedure as it establishes, in effect, dual control over individuals and decisions, with the political outweighing the economic or pragmatic.

The pervasive deleterious influence of these elements on an economic system over time leads to the elimination of common sense in decision-making, which manifests itself in low grade information, planning without facts, and distorted evaluations. Corrective mechanisms are destroyed and decision makers inhibited and rendered incompetent in the face of an absurd complexity of procedures and rigidity of targets. Given these fundamental negative conditions, the customary package of economic reforms was irrelevant. For Drewnowski, the question was not one of material incentives, or changing the structure of the economy, or modifying the principles of planning, or introducing a degree of decentralization. He insisted that an effective remedy for this endemic problem could be found only in the restoration of basic human freedoms and democratic rights. Hence the democratization process begun by Solidarity was the only possible basis for the effective improvement of the Polish economy, because the reforms it called could address over the long run the deterioration of the country's economic tissue (Drewnowski, 1982).

Drewnowski's diagnosis most closely fits Stalinist socio-economic forms at their peak, when the elements of tissue degradation are at their most invidious. The fact that classical Stalinism as a political and economic style had long passed its prime in Eastern Europe did not, however, invalidate the usefulness of his analysis. The structures of Stalinism remained, albeit in faded form, to affect the centrally planned economies of the region.

Ziegler (1987) has noted that in pluralist societies, responsibility for environmental protection is spread throughout the political system, with individuals, private corporations, government officials, and agencies all involved with and at times held responsible for this function. These actors may suggest structural changes in the economy or society that challenge basic assumptions. By contrast, in Soviet corporatist political-economic systems, responsibility for the environment must flow through the Party-State structure (Ziegler, 1987).

Morawska (1982) identified several reasons for the lack of functional pluralism in Poland. First was absence of an adequate standard of living, an element without which pluralism is virtually impossible in any country. In

Poland, to be sure, an adequate standard of living was a relatively recent and transitory phenomenon. The economic boom of the 1970s had for a time heightened social and personal economic aspirations. Morawska noted that its collapse and the ensuing general sense of economic insecurity did little to foster an environment favorable to the development of working social and political pluralism (Morawska, 1988).

Against this social backdrop operated another factor: rule by a minority party which continued to try -- unsuccessfully -- to control all important political activities. Largely as a result of an unabated lack of legitimacy, the Polish United Workers' Party (Polska Zjednoczona Partia Robotniczna, or PZPR) had been unable or unwilling to eliminate the network of unofficial political, educational, and publishing organizations that consituted Poland's parallel political structure. The PZPR was able to prevent these socio-political networks from realizing their demands yet was itself prevented by them from achieving its goals. The PZPR could not permit the establishment of the straightforward cure for this situation -- political pluralism -- for to do so would have resulted in its being swept out of office (Morawska, 1988).

Finally, the absence of political pluralism was assured by the lack of comprehensive economic pluralism in the country's centrally planned economic system. Despite continual efforts to operate its centralized enterprises efficiently and productively, the regime's failure to do so led repeatedly to major production failures and mounting economic crisis that sparked four popular uprisings over twenty five years in 1956, 1970, 1976, and 1980 (Brus, 1983). In the face of unsustainable pressure from Solidarity, the government responded to economic and political crisis by adopting a program of "pluralist" market socialism aimed at loosening the centralized control of economic enterprise. Despite several years of effort in this direction, the far from encouraging result was a situation, which took shape in the 1980s, in which some market-based elements were added to the structure of centrally regulated enterprises. This served only to perpetuate the economic crisis that was made (barely) manageable only by the existence of a thorough-going informal or second economy (Los, 1987).

It is instructive that an assessment of Poland's energy-based environmental problems prepared in the mid-1980s read like a prescription to correct this political situation. Poland's economy, like those of its neighbors in the former Council for Mutual Economic Assistance (CMEA), uses on average two to three times as much energy and raw materials per unit of generated national income as the economies of the industrialized West. The environmental effect of this profligacy is the economic root of Poland's ecological crisis.

Addressing this question in 1984, Dr. Wlodzimierz Bojarski of the Polish Academy of Sciences came to what can only be called remarkable conclusions for a technical paper when he asserted that the Polish economy was being choked not by quantitative deficiencies, but anomalies of a qualitative nature (Bojarski, 1986). Put into contemporary Western terms, the product line of the Polish economy as a whole showed a despairingly poor ratio of information to mass in its output. Its products required excessive inputs of energy and raw materials whose use continues to place an unacceptable burden on the country's environment and resource base but whose quality and innovation content was so low as to virtually obviate the idea of value added. What is significant is that Bojarski, a technical expert, did not merely propose technical solutions to the problems thus posed -- acknowledging that they too were required -- but moved in his analysis immediately to an exploration of the social and political changes that had to come about in the country's economic system if technical progress were ever to take place.

Bojarski (1986) maintained that the problems of Poland's energy policy were squarely institutional, with the only remedy nothing less than a healing of the country's "economic tissue," badly abused by forty years of economic planning that was sometimes mistaken, sometimes corrupt, but at all times imbued with the melancholy attributes of Stalinist-style management: centralized control, and achievement of quantitative objectives (Drewnowski, 1984; Bojarski, 1986). The brake on the country's economy was the over-centralized, plodding, conservative process of central planning, with its prediliction toward extensive, large-scale approaches to economic investment and rigid financial and allocational structures that could not accommodate inter-sectoral and multi-sectoral programming. This system effectively stymied the development of the requisite number of adaptable, flexible, and efficient small and medium sized firms that could operate on a small scale all over the country on the initiative of workers and communities themselves (Bojarski, 1986).

## Environmental Affairs in the Period of Transition

Poland, Bojarski said, was facing the prospect of the collapse of the regenerative capacity of its natural environment and the exhaustion of its natural resources that would result in nothing less than the collapse of its economy and social order. This prospect could be staved off only by measures on the part of the Party and the government that tapped the social energies of cooperation to rebuild the country's social and political system (Bojarski, 1986). This rebuilding had to include such far-reaching elements as currency reform, elimination of economic monopolies, the removal of politics from economics, and the restoration of openness in Polish political life. Needless to say, such ideas, which contained within them no small measure of the reweaving of civil society, were viewed in some official quarters as quite threatening. None of this, Bojarski insisted, made for the return of capitalism, for it never proposed handing over major industrial enterprises and financial institutions to private owners, abdication of the processes of economic planning and state intervention, or a retreat from pursuit of social goals. It was, in fact, not greatly different from the Hungarian economic reforms or the reforms undertaken in China in the 1980s. It was, as Bojarski stressed, a program entirely within the bounds of the Polish constitutional system, in fact more so than the country's socio-economic system at the time -- a system so reliant on irregular and extraconstitutional intervention (Bojarski, 1986).

This far-reaching program, first proposed in this form in 1984, parallelled to a significant degree the theses of the government's own Second Stage of Economic Reform put into effect in 1988 despite their failure in a national referendum in the prior year. Whether these market-oriented reforms would have been able, as their framers hoped, to salvage Poland's economy within the context of the country's communist political system is something never to be learned. Reality moved too fast and too surprisingly to allow the reforms their time. Curiously, the policy content of today's transition to a market economy is not markedly different. The critical difference is the wholesale change of the political background, which was the crucial variable all along. For purposes of the present analysis, the critical aspect is that these reforms were proposed for the rather different purpose of rescuing Poland's economy from its overwhelming and damaging energy and material intensiveness.

Environmental concern arose in Poland, as it does everywhere, as a social response to a perceived threat. Social responses to such threats take the form of change or creation of institutions. The threat posed by environmental degradation was responsible for the emergence of new organizations and the restructuring of old in both the official, or public, sector in Poland and in the emerging private sector of society. In the official sphere, this included the expansion of governmental environmental agencies, the creation of state-sponsored social organizations addressed to environmental goals, and the expansion of the role of quasi-governmental institutions such as the Academy of Sciences. Similarly, in the private sphere, this process comprised the creation of independent environmental groups of varying political persuasions and the redefinition of fairly conventional functions such as the role of research programs in universities.

In Poland, the stimulus to institutional change was the generally agreed upon failure of environmental policy. This failure gave rise to responses that took their direction from different elements of society: the deep desire in

Polish society for authentic political structures and practices, the government's effort to retain legitimacy, the evolving role of the scientific professions as sources of information for society, and the anti-regime political premises of opposition organizations that often underlay concern for environmental quality. It is this mix of sometimes competing, sometimes complementary efforts which characterized the decline in the 1980s of monolithic decision-making in Poland's political system. Nowhere were the symptoms of this decline and those of incipient political pluralism more evident than in the arena of environmental affairs. Poland's nominally communist-dominated polity was transformed by the emergence of a range of environmental interest groups that sprung up to parallel and challenge the country's governmental bodies and officially sanctioned organizations.

The absence of political pluralism in Poland acted as a key factor in the largely undisturbed growth of ecologically deleterious practices in all sectors of the economy and society. As long as the communist regime was able to maintain substantial political control -- by means that excluded authentic political participation -- the de facto absence or weakness of feedback mechanisms in decision processes inhibited the incorporation of environmental issues into the political agenda. At the same time, when these issues had clearly come to the attention of the Polish public, the regime found itself unable to use them to legitimize its rule. In fact, environmental issues came to be a *delegitimizing* tool wielded by the country's political opposition. A sort of symbiosis between environmental activism and the political pluralism took hold in Polish society over the 1980s. Incipient pluralist forms implanted themselves in Polish political soil, and environmental affairs anchored themselves in this changing social context of growing pluralist-style political participation. At the same time, exercise of the possibilities of participation by an environmentally concerned public served as a testing ground for public political participation overall. The enduring impact of the brief Solidarity period was the demonstration it provided to the people of Poland that the management (and implied reorganization) of their society -- so long formally monopolized by the communist party -- was entirely within their competence. To this should be added the unstopping of the free flow of information that Solidarity bequeathed to Polish society. Information played a crucial role in environmental protection affairs in Poland. The knowledge of how things worked -- learned during the Solidarity era -- made it possible for Polish society to face up to (and, were it not for the Soviet factor, do away with) the Polish regime. The technique of using information to force issues onto the public agenda made the Polish environmental movement possible by forcing the government to acknowledge the reality of the issues and then take action in order to justify its behavior. Information -- to propose, to confront -- was the only weapon which the ecological opposition had with which to challenge the plans of the heavy industry lobby. While public opinion was not the lever it is in other political systems, even communist regimes had begun to find that they had to keep abreast of popular dissatisfaction or risk losing control of events in their countries.

In the late 1980s the question remained whether moderate public knowledge of environmental issues constituted an aspect of openness in the Polish polity or merely served as grounds for further frustration with regime inaction. That is, did greater public information allow the average Polish reader to conclude that problems discussed might become problems addressed, or did the apparent public acknowledgment of problems unmatched by indications of forward movement merely lead to further popular disenchantment? Did communication among scientists constitute a practical basis for influencing environmental and development policy in positive ways, or had official inattention and lack of broad dissemination rendered their work, in the worst sense, academic?

After so many years of obfuscation, withholding of information, and outright misrepresentation on so many subjects, the Polish regime suffered from the syndrome of the boy who cried wolf. It had been caught making misrepresentations and telling outright lies so many times in the past, that even when it told the truth, the people were inclined not to believe it. In the environmental arena, as in virtually all spheres of civil life in Poland, the opposition had the ear and the trust of the populace. A combination of well-developed reflex and calculated self-defense caused people to believe unofficial sources and discount the government's reports. Ironically, it seemed that there was one circumstance where the credibility of government information was not minimized but rather magnified: when the government reported that a given situation was bad, Poles were inclined to conclude that it was even worse than reported.

To be sure, this was not a recipe for successful social mobilization in a context where to address the pressing needs of environmental protection might mean diversion of economic resources that would constitute further infringement on the standard of living in a country where economic austerity had followed economic crisis to encroach on the average citizen's financial means. In such a situation political good faith -- if not full trust -- was critical but sadly lacking. In 1987 the Polish Academy of Sciences asserted that to continue along this path would lead Poland to social catastrophe stemming from the political stratification of society. At the time, this was a reasoned and scholarly way of putting the problem that was described less circumspectly this way by another Polish observer of the scene.

It is hard to imagine environmental degradation not playing a role in the next Polish crisis or to imagine that this crisis is very far off. People are aware of the degree of ecological degradation and the threat it poses to them and their tolerance will not hold for long. (Sitnicki, 1988)

As these remarks indicate, despite increasing activity in all quarters, the field of environmental affairs in Poland was not without the potential for conflict and tension. To a great degree this tension mirrored conflicts in society as a whole that went far beyond the obvious antagonisms between those whose first interest is the environment and those who continue to place other interests, such as industrial investment, first.

Poland's ecological movement was born as a social protest against the growing threat to the natural environment, made possible by the removal in 1980 of the blockade on information in Polish society. This remarkable emergence of private organizations and interest groups concerned with environmental issues flies in the face of the model of society adhered to by communist theoreticians in which information and social organization are controlled by a communist party which acts in the interest of the people. The existence of these groups defines an alternative and competing model that may be termed incipient political pluralism. An outline takes shape of a society in which self-starting groups of like-minded citizens organize and coordinate actions in pursuit of avowed goals as independent actors fully capable of functioning on their own. A frame of reference for this type of behavior had existed in Polish society since 1981, but its operation was largely restricted to areas outside official politics. The question answered by the events of 1989 was: when would this vigorous stream of political power overwhelm the barriers keeping it below the surface of society and break through to reconstitute the nature of the mainstream.

The decentralizing thrust of the 1988 Second Stage of Economic Reform was clearly an attempt to introduce elements of initiative and flexibility into Poland's calcified economic system. It was based on the assumption that broadening responsibility for economic decisions would engender parallel responsibility for the economic system as a whole. Seen this way, the reforms had as their long-term object that ever elusive goal of the country's communist regime--the achievement of legitimacy in the eyes of the people. While certain communist regimes had been able to achieve a degree of legitimacy without allowing popular participation in politics, Poland's elite was reduced to striving for the former by means of the latter.

As a result, there was the danger, when viewed from the former Party headquarters at the corner of Jerozolimskie and Nowy Swiat in Warsaw, that participation in economic decision making would carry over into participation in political decision making of a sort -- as Poles were wont to make it -- unacceptable to a Party ever jealous of its ideologically mandated leading role in political affairs. Poland's recurrent political crises regularly took on forms quite different from the specific events that triggered them, often exceeding the expectations of both sympathizers and antagonists. It was entirely possible that restive Poles, ostensibly moved to action by an intolerable ecological threat, might be found giving vent to a decade of internalized dissatisfaction with matters social and economic and once again actively calling into question their country's entire political and economic system (Kabala, 1991). As reality became in 1989 even more remarkable than fiction, Poland moved precisely along these lines. Since then, Poland's political revolution has taken away the center-stage role of environmentalism as an engine of political change and submerged it in a sea of difficult, competing social issues.

While this view concludes at a watershed of Polish (and European and world) history whose impact on all policy making and politics in Poland was revolutionary, it is averred that the patterns, habits, structures, and practices which took shape in this era will continue to influence unfolding events in Poland for some time to come.

# References

Bojarski, W. 1986. Ways of Overcoming the Economic, Energy, and Ecological Barriers to the Development of the Country. In *Energetyka, Srodowisko, i Planowania Rozwoju (Energy Policy, the Environment, and Development Planning)*. Warsaw: Polish Academy of Sciences Committee on Environmental Engineering and Institute for Basic Technical Problems.

Brus, W. 1983. Economics and Politics: The Fatal Link. In A. Brumberg (Ed.), *Poland: Genesis of a Revolution*. New York: Random House.

Drewnowski, J. 1982. The Anatomy of Failure in Soviet-Type Systems. In J. Drewnowski (Ed.), *Crisis in the East European Economy*. London: Croom and Helm.

Fallenbuchl, Z. M. 1965. Some Structural Aspects of the Soviet-Type Investment Policy. *Soviet Studies, 4*.

Fura, Z. 1985. The Polish Ecological Club. *Environment, 11*.

Kabala, S. J. 1985. Poland: Facing the Hidden Costs of Development. *Environment, 11*.

Kabala, S. J. 1991. The Development of Environmental Policy in the Polish People's Republic, 1969-1989. (Doctoral dissertation, University of Pittsburgh, 1991.)

Kassenberg, A. 1990. Diagnosis of Environmental Protection Problems in Poland. In W. Grodzinski, E. B. Cowling, & A. Breymeyer (Eds.), *Ecological Risks: Perspectives from Poland and the United States*. Washington, D.C.: National Academy of Sciences.

Kassenberg, A., & Rolewicz, C. 1985. *Przestrzenna Diagnoza Ochrony Srodowiska w Polsce (The Spatial Diagnosis of Environmental Protection in Poland)*. Warsaw: Polish Academy of Sciences Committee on the Spatial Development of the Country.

Kormondy, E. J. 1980. Environmental Protection in Hungary and Poland. *Environment, 12.*

Kostrowicki, A. S. 1983. Straty Ekonomiczne Wynikajace z Degradacji Srodowiska (Economic Losses Resulting From the Degradation of the Environment). In A. Ginsberg-Gebert (Ed.), *Ekonomiczne Problemy Ochrony Srodowiska (Economic Problems of Environmental Protection)*. Warsaw: League for the Protection of Nature.

Lange, O. 1983. *On the Economic Theory of Socialism*. Minneapolis: University of Minnesota Press.

Los, M. 1987. The Double Structure of Communist Societies. *Contemporary Crises, 11.*

Misztal, B. A., & Misztal, B. 1984. The Transformation of Political Elites. In J. Bielasiak and M. Simon (Eds.), *Polish Politics: The Edge of the Abyss*. Boulder: Westview Press.

Morawska, E. 1988. On Barriers to Pluralism in Poland. *Slavic Review.*

Pudlis, E. 1983. Life Without Fluoride: The Continuing Story of Polish Pollution. *Ambio, 6.*

Sadowski, C. M. 1981. Citizens, Voluntary Associations, and the Policy Process. In M.D. Simon & R.E. Kanet (Eds.), *Background to Crisis: Policy and Politics in Gierek's Poland*. Boulder: Westview Press.

Sitnicki, S. 1988. Remarks to author. Central School of Planning and Statistics, Warsaw.

Staniszkis, J. 1983. Solidarnosc Jako Zwiazek Zawodowy i Ruch Spoleczny (Solidarity as a Trade Union and Social Movement), in W. Morawski (Ed.), *Demokracja i Gospodarka*. Warsaw: University of Warsaw Institute of Sociology.

Symonowicz, A. 1988. Straty z Tytulu Degradacji Srodowiska: ich Charakterystyka i Probu Szacunku (Losses from the Degradation of the Environment: Their Character and an Attempt at Estimation). In A. Ginsberg-Gebert & Z. Bochniarz (Eds.), *Ekonomiczne Problemy Ochrony Srodowiska (Economic Problems of Environmental Protection)*. Wroclaw: Ossolineum.

Wright, A. W. 1971. Environmental Disruption and Economic Systems. *The ASTE Bulletin, 1.*

Ziegler, C. 1987. *Environmental Policy in the USSR*. Amherst: University of Massachusetts Press.

Chapter 4

# ENVIRONMENTAL POLICY IN A TRANSITION ECONOMY: THE BULGARIAN EXAMPLE

Kristalina Georgieva

The on-going "marketization" in the former socialist countries, triggered with enthusiasm and followed by growing disappointment and pessimism, clearly dominates all social processes in the region. Its implementation is often badly managed, which leads to chaotic adjustment in the affected social subsystems (industries, services, etc.). Particularly in the field of environmental policy design, efforts to radically upgrade the priority of environmental protection and integrate it with the mainstream of the reform process have been limited -- despite the serious pollution problems facing the new governments and despite the popular environmental movements.

This is especially worrisome in the case of Bulgaria, where environmental concerns provided the common ground for antitotalitarian opposition in the past. Although the achievement of an "ecological market economy" was -- and still is -- proclaimed, after the first several months' euphoria of democratic governing the role of environmental policy-making continues to be marginal and conducted separately from major economic changes. In his address to the Parliament on March 23, 1992, the Bulgarian Prime Minister Philip Dimitrov (former vice-chairman of the Green Party, now member of the Conservative Ecological Party) dropped environmental issues to the bottom of his first non-communist government priority list and left an impression that, aside from introducing the European Community's standards (a highly disputable task), no serious environmental policy steps were considered.[1]

How do we explain and, moreover, how do we deal with the phenomenon of insufficient environmental friendliness of democratization? In the search for an answer, this paper explores the problems of environmental policy adjustment in a postsocialist country during the transition from a centrally planned to a market economy. Its primary goal is to evaluate the experience in environmental decision-making accumulated since the beginning of transition, to indicate the most serious constraints to integrating environmental considerations within the economic reform, and to make suggestions for further policy developments.

Describing the historical *background* for environmental policy-reshaping in Bulgaria aims to present the heritage from socialism, and also to point out that despite the weaving of environmental problems into the economic

*A. Vari and P. Tamas (eds.), Environment and Democratic Transition, 67–87.*
© 1993 *Kluwer Academic Publishers. Printed in the Netherlands.*

crisis of Central and Eastern Europe, key steps in the reform process there are being taken without regard to possible environmental consequences.

In addition, some *constraints* to quicker and more effective replacement of the system for environmental protection and management inherited from socialism are analyzed and some ideas to overcome them are suggested.

As the financial constraint to fundamental changes in the environmental sector is the most severe, the paper further focuses on the possibilities for introducing some *cost-saving incentives* for more rational business and consumer behavior both from environmental and from economic point of view.

Although only the Bulgarian experience is closely investigated, most of the indicated problems and suggested solutions are valid for other European postsocialist countries. Differences in size, level of economic development, and speed of reform among these countries, as well as in support from governments, businesses, and international institutions in the developed world, lead to variations in the countries' adjustment capability, including on environmental issues.

The turbulence of transition makes all attempts to predict the best options for further environmental policy development extremely difficult. The post-totalitarian countries are entering a highly complex adjustment process, within which they have to deal with a huge set of problems: political, economic, environmental, psychological, ethical, and so forth. To decrease the corresponding uncertainty, some questionable assumptions are made; for instance, that political and economic stabilization will be successful; there will be nation-wide acceptance of restitution of capitalism; rapid development of market infrastructure; speedy recovery of key economic branches, transnationalization of the economy, etc. We are going to be modest optimists, who believe that by mid-nineties all major goals of transition will be accomplished, and therefore it is worth building an environmental policy framework for the then "normal" market economy. Still, the "What if not?" question, here repressed, does exist, and raises well grounded fears about the success of environmental policy adjustment, and about the future of the environment itself.

## Environmental Policy Background

### Preliminary Remarks

Prescriptions for environmental policy changes can hardly be written outside a comprehensive "state-of-the-art" picture, which describes both the current condition of the environment and the policy steps undertaken to improve it. For the purposes of this paper, only one dimension is included -- the environmental policy heritage. Evaluations of the state of the environmental sector have already been presented in two Bulgarian *State of Environment Yearbook* (published in 1991 and in 1992) and in other official publications.[2] Like the rest of the former Eastern bloc, Bulgaria is now experiencing the consequences of past environmental neglect. This straightforward statement requires some additional comments.

First, although the environmental situation in Bulgaria deteriorated severely during the last decades, it is in relatively better shape compared to other Central European former socialist countries (the former East Germany, Czechoslovakia, and Poland). An explanation can be found in the Bulgarian pattern of "socialist industrialization," in which heavy industries played a less dominant role. This pattern is determined by: (i) historical trends (Bulgaria began a capitalist development in the beginning of this century and emphasized agriculture and light industries); (ii) scarcity of basic natural resources for industrialization (coal and iron ore were the engine of socialist industry in Poland, Czechoslovakia, and the former East Germany); and (iii) ambitious high tech development programs during the 1970s, which did not succeed in promoting efficiency and competitiveness, but left a heritage of lower pollution.

Second, the scale of environmental degradation differs not only from one country to another, but also among different regions in a given country. The level of pollution is much higher in particular areas (so-called "hot spots"), whereas other zones are relatively clean. These differences are due to three main reasons:

· The socialist version of the NIMBY (Not In My Back Yard) problem, which in the former socialist countries existed in the peculiar form of *NIOLBY* (Not In Our Leaders' Back Yards) -- a matter of prestige for a high level Communist Party official was to prevent his or her native village/city from "dirty industrialization;"
· The regional differentials created by the *Cold War*, as zones near national borders were kept less industrialized. As a result, they experienced high outmigration, and are now facing serious problems in recruiting population, especially young families. At the same time these areas are relatively clean, a factor for comparative advantage in their future economic development which is seldom considered;
· *Authoritarian decision-making*, which geographically allocated large industrial plants during industrialization, without regard for potential pollution problems or the absorption capacity of local environments.[3]

Generalizations on regional or on country level, both in exaggerating or underestimating the severity of existing environmental problems, are risky and should be avoided. From an economic point of view too much pessimism could be as inefficient as the previously overemphasized socialist optimism. The transition bears a high price, and scarcity of investments imposes a severe constraint to environmental spending. Exaggerated claims about the level of environmental degradation are often presented by green activists both from the East and the West in a noble attempt to draw more attention to the state of the environment in the region. Instead of achieving this goal, they often have a negative effect by misleading policy-makers and creating negative investment reactions abroad (fears of a too high level of inherited pollution; preferences for investments in dirty industries; threats to the unpolluted areas, etc). This attitude is especially unfavorable for the otherwise relatively good future of the Bulgarian tourist industry. Nonetheless, one of the emerging paradoxes of transition is the tendency to jump from one extreme to another by substituting the environmental optimism of "real" socialism with complete environmental pessimism of postsocialism. Eastern and Central European (including Bulgarian) environmental activists, academics, and politicians tend to describe the situation worse than it is, supposedly in an attempt to support a "green restructuring," but mostly expressing the still strong anticommunist criticism.

With this call for realism also comes a requirement for further work to verify the interdependence of the economic and the environmental sectors -- a rather difficult task given the necessity of complete rearrangement of the system of national accounts, recalculation of data, and raising the level of accuracy of both economic and environmental information.[4]

*Environmental Policy in Retrospect: What Was Done During Socialism?*

As early as the 1950s and 1960s, attempts to develop environmental policies were initiated in the then socialist countries in response to a worsening environmental situation at home and growing attention to the environment in the developed market economies. Bulgaria was not an exception. An already existing environmental law (the first *Nature Protection Act* was issued in 1936)[5] was further expanded in *Nature Protection Act of 1967.*[6] Some additional laws were passed, directly or indirectly covering different environmental resources (*Mining Act* of 1957, *Forestry Act* of 1958, *Water Act* of 1969, *Land Protection Act* of 1973, *Hunting Act* and *Fishery Act* of 1982, *Sea Act* of 1987).[7]

In accordance with the environmental laws, State *Decrees* were issued setting stringent ambient and/or effluent standards for air and water pollutants, waste disposal rules, and land use norms. The standards are strict, in some cases more strict than those in the developed countries. Unfortunately, they are also demand more detailed data than the monitoring system can provide. For example there are 87 water quality standards, but measurement is technically possible for only 25 of them, and no more than 6-8 are regularly controlled.[8]  Similar problems exist in regulating air pollution, land use, waste disposal, and noise.

The environmental policy framework was completed with the formation of state and regional institutions (Committee for Environmental Protection and Regional Environmental Protection Inspectorates).  In 1975, environmental protection was incorporated in the planning process. A new section was added to the state five year plan, dealing with pollution prevention and control.

The most important feature of this legislative and regulatory framework is that it allowed and (supposedly) targeted a very strict environmental command-and-control policy.  In fact such policy has never been implemented, because: (i) pollution was legally tolerated in the case of high state priority production; (ii) enforcement was not guaranteed; (iii) charges and fines for non-compliance were set below the costs of compliance in almost all cases; and (iv) the Committee for Environmental Protection and the Regional Inspectorates operated with limited budget and partocratic apparatus.  The managerial capability of environmental bureaucrats is on average worse than in most other administrations due to comparatively low empowerment and low priority of their functions.  Central planning failed to produce a working command-and-control system.  The apparently ideal composition of legislation, standards, institutions, and economic measures (charges and fines) has been, and still is, a display for the outside world.

At the local level, the meaninglessness of laws vis-a-vis party instructions, and avoidability of national requirements leaves deep distortions in polluters' patterns of behavior.  The inherited deficiency of the enforcement system reflects weaknesses in the institutional mechanism.  Local officials, who know well the state of the environment, are either powerless or restricted in fulfilling their controlling functions because of their relations with local factory managers or administrators (friendships, "connections", past or future interdependence etc).

The dynamics and the structure of environmental spending also elucidate the shortcomings of the regulatory system.  Since 1975 the share of environmental expenditures gradually rose, reaching a peak of 1.8 percent of the national income in 1987,[9] and afterward declining to about 1.2 percent in 1990.[10] The distribution and use of scarce financing were also very inefficient, mainly because priorities were set on political grounds, and investment programs were systematically delayed.

A major obstacle to effectively tackling the country's deepening environmental problems resulted from the lack of nation-wide environmental strategy.  The composition of environmental spending (for air, water, land, noise, protected territories, and recycling technologies) had been fluctuating, mostly because of shifting funds from one area to another in crisis response to sharply deteriorating environmental conditions.

Attempts to promote prevention versus cleaning-up environmental policies have not been very successful, especially in introducing the so-called "waste-free technologies" (technological processes with high level of raw materials utilization, water and waste recycling). Their targeted share of 35 percent of the environmental expenditure has never been achieved.

The overall trend in the dynamics of these expenditures is hidden behind the availability of only current price statistics, which leaves the impression that during socialism environmental spending registered a high and stable rate of growth. In fact, the growth rate has been negative since the end of 1980s if constant prices are used. For example, in 1989 a 40 percent average increase in prices concealed the decrease in almost all types of environmental expenditures. The use of current, rather than constant, price statistics creates serious difficulties for objective analysis, particularly in the case of measuring environmental costs and benefits, especially since the liberalization of prices and exchange rates in the beginning of 1991.

Besides being inadequate, money was also inefficiently spent, and only 55-65 percent of the planned investments were actually made. Equipment, supposedly installed, was kept in back yards, as management preferred reaching the production quotas to meeting pollution standards. Project delays and postponements were (and still are) very common.

In short, at the starting point of transition, environmental policy in Bulgaria, as elsewhere in the former Eastern bloc, had to deal with the consequences of:

(1) Rapid industrialization, combined with extensive economic growth (*"rush growth"*), and high natural resources intensity of production;

(2) Free or underpriced consumption of environmental resources, considered "people's property";

(3) Limited application of economic incentives, and soft financial constraints on waste and pollution (low raw materials prices; low or free of charge waste disposal services; equally soft credit conditions for all producers, including those generating excessive waste and pollution; low level of pollution fines, etc.);

(4) Hypercentralization of decision-making, leading to inefficient resource allocation and use;

(5) Underinvestment in environmental protection and systematic project postponements and delays;

(6) Inherited Soviet technology, designed for low-population-density regions, transferred to high density regions in the smaller Eastern European countries;

(7) Information scarcity (both in terms of quantity and quality of data);

(8) Low social priority of environment, and managerial neglect to environmental problems; and

(9) Communist ideology constraints to sound environmental policy -- i. e. low-tech industrial development, which favors the proletariat (services were considered "non-productive"); overambitious planning goals; transfer price promotion of targeted economic branches -- combined with bureaucracy and corruption.

All of the above factors are going to have medium or long-term effects on the state of environment, as well as on environmental policy in the transition to a market economy. However, their elimination needs the implementation of three different sets of solutions. Organizational and institutional changes are mainly required for the switch from a centrally-planned to a market economy (1 to 3) with indirect effects on 7, 8, and 9; economic restructuring -- for the abolishment of the Soviet industrial model (4 to 6); and behavioral, socio-cultural and educational changes -- for both changing perceptions of the environment and adjustment to capitalism (7, 8, and 9).

## Entering the Transition Period: Intentions Versus Reality

In the process of "marketization", the establishment of legislative, institutional and regulatory environmental infrastructure is crucial. It is a prerequisite to further structural changes and to human resources adjustment. It can bring results in short run and has an important international impact on reducing the gap between the East and the West. The promotion of legal and institutional changes has been systematically included in the list of recommendations by international institutions, Western governments officials, and environmental policy experts. Since 1990 the Bulgarian environmental policy-makers have been promoting an ambitious new agenda, in line with the requirements for changes. An evaluation of their progress so far can help us formulate some general conclusions about the nature and the speed of environmental policy adjustment under transition.

The clearest phenomenon of the first two post-totalitarian years (1990 and 1991) is the discrepancy between the stated high environmental policy targets and the actual low interest in immediate action, penetrating into legislative activities. Although the society was assured that a leading priority for the new legislators was saving the environment, in fact very little work was devoted to integrating environmental concerns in the transitional policy framework.

The first and the second communist (socialist) dominated post-totalitarian governments followed the past tradition of wishful thinking by developing an ambitious, completely unrealistic, and not related to legal and institutional changes strategy for environmental improvements. In

1990 an emergency program for restoration and protection of the environment was announced. It included 134 well defined emergency measures, targeting the most dangerous sources of air and water pollution, land degradation and contamination, radiation, and endangered species. Despite being well tailored to the existing needs, the program suffered the typical weakness of socialist wishful thinking as only about 25 percent of the suggested actions were financially and organizationally supported.[11] Another program for water quality declared that all rivers should meet the requirements for at least II class (i.e. average level of pollution). Considering the state of water resources black-humorous environmentalists suggested that the only clean river Veleka had a chance of achieving this goal.

The visible goal of this strategy was to set targets too high for the non-communist inheritors and by that, to predetermine their policy failures. As in other areas, this approach proves to be fruitful, and slowly turns part of the public to the left (pro-socialist) again -- an unfortunate, but relevant point in predicting the most probable social (including environmental) policy options in a medium and long run.

The key environmental legislation -- a fully revised Environmental Protection Act (EPA)[12]-- was worked out in the shadow of political struggles. The slight socialist majority in the first post-totalitarian Parliament turned almost all legislative incentives in battles for changing the power balance in favor of the opposition. As a result more time was spent on procedural questions, personal accusations, and political speeches, then on conceptual discussions. Unfortunately, the EPA was among the worst affected legislations. The EPA targets the environmental policy adjustment to the requirements of the transition period, and the future market economic organization. The project was prepared mostly by parliamentarists from Ecoglasnost, and after being introduced to the Parliament it became more a subject for inter-party trade-offs than for serious legislative discussions. Hurriedly passed days before disbanding the Parliament, the Act is a good illustration of both the positive and the negative tendencies in the policy field, which in one way or another now take place in all former socialist countries. The second post-totalitarian Parliament, elected in 1991, decided to reconsider some definitions in EPA. However, political tensions again put pressure on the legislators, and led to postponement of the parliamentary debates on the Amendments to EPA, as well as to controversial solutions to some issues (among them is design of the environmental fund, which is too broadly defined from conceptual and managerial points of view).[13]

The new legislation is noteworthy for a number of reasons.

First, it introduces the *polluter-pays principle*, based upon which two measures are enacted -- charges for pollution within the standards, and substantial increase of already existing fines for pollution above the standards. These steps, adopted broadly across the former Eastern bloc, deserve both support and questioning. Very little is done so far to monetize the damage caused by different polluters, nor the full costs their activities impose on the society. Therefore, the polluter-pay-principle is more an ideological concept, then a working tool. The imposed higher charges and fines also create some doubts in its effectiveness, since no proper economic analysis supplements them, nor an explanation of how these charges and fines are going to be adjusted to inflation (especially if the inflation rate escapes regulatory control).[14] The opportunity to use taxes on polluting goods as instrument for influencing the producers through consumer behavior is not even implied.

Second, the Act legalizes the *public right-to-know*. This is also a widespread and important legislative improvement in the former Eastern bloc. Based on it, the whole attitude toward collecting, processing, and providing environmental information is now changing. This is a significant challenge, and hopefully will further stimulate active public participation in environmental policy debates in the region.15

Third, the EPA provides ground for *a revision of standards*, based on a combination of four requirements: to assure health safety; to guarantee the employment of best available technology; to reflect qualified expertise; to match the international experience. It is not clear whether all of them should be met, and when priority should be given to one requirement or another. The desire to "enter Europe", as everywhere in the former socialist countries, has led to introducing the standards of the European Community (EC) as substitutes in case Bulgarian standards are not available. Given the reality that the existing set of strict standards are not enforceable, it is doubtful if this section of the EPA would serve well the environmental policy adjustment.

Fourth, the endorsed *distribution of administrative obligations* among the main environmental institutions -- the Ministry of Environment (MOE), the environmentally related divisions of Ministry of Public Health, and Ministry of Agriculture, the Regional Environmental Inspectorates (REI), the municipal authorities -- creates expectations that the current overlap in their activities will be eliminated. More important, but unregulated, is the relationship between the national and local (regional and municipal) levels of decision-making. A major requirement for a successful environmental policy is effective decentralization, but this seems to be a highly complicated task -- opposed by the Ministry of Environment because of fear of corruption and lack of competence at the local level. These well grounded reasons are also combined with less rational factors, like the socialist inertia of centralized decision-making and attempts to keep more working places on a ministerial level.

Fifth, *environmental impact assessment (EIA) is declared mandatory for new investments*. This point still provokes discussions in two directions. Some authors critique the definition and the procedure for environmental impact assessment, which they think is likely to concentrate only on risk analysis, and to disregard low risk negative impacts.[16] A second objection reflects concerns that the Act is soft on existing sources of pollution and strict with new sources, including with respect to EIAs, which are required only from the latter. From a macroeconomic prospective this approach is highly disputable, given the limited new investment resources, and also the expectation that new investments -- if encouraged -- could bring better equipped and less polluting production units, and therefore should be encouraged.

Another debatable issue is legalizing only environmental impact assessment among many decision-motivating techniques. Other methods (cost-benefit analysis, cost effectiveness analysis, multiple criteria analysis, risk-benefit analysis, and decision analysis) also deserve legislative attention. The argument that at the moment neither benefits nor costs can be properly measured does not explain why a longer-term legislative approach is not used (addressing the future when the price distortions will be overcome and better benefit and cost estimates will become possible).

Sixth, an *earmarked fund for environmental protection* was established, mostly based on the collection of fines and charges.[17] The Environmental Fund was originally designed with a decentralized profile -- 50 percent of the total amounts is to be granted to the municipalities, 40 percent to the REI's, and 10 percent only to the MOE's Environmental Fund.[18] The adoption of this concept is highly contentious for two reasons: (i) funding is required to finance nationwide and cross-regional environmental projects (some of them already underway); and (ii) local authorities often lack either managerial experience or commitment to environmental action, and contribute to inefficient use of environmental funds. In the Amendments to the Act, the centralized share is suggested to increase to 50 percent, and the decentralized -- to merge into one only type of fund (Municipal Funds).[19] The Ministry of Environment officials argue that in the field of environmental protection decentralization should start from creating local capacity for decision-making, public awareness and involvement, and then redistribute funds from the national to the municipal level. This position is being broadly supported by non-government specialists, who either fear a high level of corruption at local level, or are troubled by diminishing environmental concerns among the population, and among local officials.

The Environmental Protection Act reflects the peripheral position of environmental policy among the new legislative initiatives. It suffers from compromises, from lack of explicitness -- especially in institutional and enforcement aspect -- and from limited promotion of economic incentives.

If we are to ask what kind of environmental policy this Act is going to promote, the answer would be: highly centralized; difficult to enforce; emphasizing end-of-the-pipe (post-pollution) regulatory activities. Neither the EPA, nor the Privatization Act define property/use rights (private, public, common) over environmental resources -- a crucial requirement for a successful environmental policy after the transitional period.

It can be argued that the reshaping of environmental policy in Bulgaria is still at a pre-transition stage. Although a strategy for environmental policy adjustment and even an action plan were written jointly by the Bulgarian Ministry of Environment and a team from the World Bank and the US Environmental Protection Agency, no substantial changes were undertaken. Instead, three naive concepts emerged at this stage: (i) that the West wants the former Eastern bloc clean and will pay for it; (ii) in case it refuses to do so, the economies in the region should target "dirty growth", and not be concerned about the environment until getting rich; and (iii) once the market economy develops, it will somehow put in action its famous invisible hand to conduct a better environmental behavior. In current environmental policy-making, preference is being given to emergency responses to the most critical problems, with little effort toward integrating environmental concerns in the mainstream of the economic reform.

## Constraints On Environmental Policy Adjustment

Let us try to determine what creates this mentality and then suggest some ways to deal with it.

### Political Turbulence and the Role of the "Greens"

Irreversible political changes are leading Bulgaria (as the rest of the former Eastern bloc) to a Western style multi-party democracy. The transformation process, not smooth in any of the transition societies, is particularly controversial in the Bulgarian case. Two reasons are of importance: lack of democratic traditions, and shortage of experienced opposition during the communist period. Many reform efforts are distracted by the political struggle; even the environmentalists are more concerned with issues of division of power than with environmental problems.

At the same time, inside and outside Central and Eastern Europe there are expectations that "green politics" are capable of playing a leading role in regional democratization and in reshaping the environmental policy. This is particularly true for Bulgaria, because the environmental movement there holds a core place in the newly born opposition both before the changes in the end of 1989, and immediately after them. The latest developments do not confirm the expected political "greening." Instead of gaining positions, the environmentalists have been losing them, and their influence is persistently diminishing.[20] Recognizing the former communist governments' responsibility for the current state of environment, and making it public is an important, but already exhausted social role for the "greens." Their willingness to take a stake in the environmental policy reform by seeking political promotions to key government positions unfavorably contrasts to their reduced efforts to encourage public participation in the decision-making process, and causes a loss of credibility. At the same time there is no sign of environmental concerns spreading over the rest of the political horizon.

Under the pressure of the worsening economic situation, the public is becoming less and less supportive of environmental incentives if they are perceived to have negative or even neutral economic effect.

There are at least three major challenges to be met in order to create more favorable political conditions for successful environmental policy.

First, the green movement needs to reshape its agenda and to face the key issues of the post-totalitarian society (such as restitution and privatization, regulatory reform, economic restructuring, growing consumerism, internationalization -- and their potential environmental impacts). Many green activists are preoccupied with political struggles (against the former communists and for post-communist distribution of power). For some opposing the current non-communist governments is difficult (because of previously built political alliances).

Second, the most important "green" issues should appear on the agendas of non-"green" political and governing entities. The necessity to promote environmental awareness outside the green movement is now broadly recognized, but still very little efforts within the green movements and the environmental institutions are devoted to outreach the important policy agents (major political parties, the trade-unions, the key reform Ministries -- of Finances and of Industry).

Third, a nationwide dialogue between parties, government institutions, trade-unions, and non-government organizations (NGOs) on the key reform issues has to be initiated, focusing on the ecological sustainability of the transition period. Facilitating a consensus through this dialogue is an important role for both environmental officials and NGOs, and a major requirement for environmentally sound economic transformation.

## Macroeconomic Disequilibrium

The most important constraint to environmental policy development is the severe economic crisis which Bulgaria is experiencing, like the rest of the former socialist countries. It is combined with dramatic changes in the pattern of social and economic life at all levels. On a macroeconomic level, a sharp decrease in production of all sectors (especially of heavy industries) and fast increase in unemployment are combined with rapid inflation, stagnated domestic and foreign demand, and fluctuating exchange rate. The stabilization policy, designed to diminish the adjustment shock, is failing to meet its targets.[21]

The neoclassical approach to restoring macroeconomic equilibrium concentrates on reducing the internal and external disbalances. Both investments and consumption are now restricted, and the shock of transformation from soft to tight budget constraint goes through the whole economy. Rapidly growing unemployment creates difficulties for managing government expenditures. In fact, all gains from subsidies' reduction are lost in rising social costs and the budget deficit climbs up. Under these circumstances the economy is facing a longer-run macroeconomic disequilibrium, and the environmental projects are the among the first to be severely hit.

Another aspect of the macroeconomic equilibrium problem, often forgotten, concerns the necessity of drastically changing the equilibrium concept itself.[22] In the western macroeconomics the macro-equilibrium usually is presented by the following equation:

$$GNP = I + C + G, \text{ where}$$

- $GNP$ is Gross National Product;
- $I$ is gross private investments;
- $C$ is consumption of the individuals; and
- $G$ is government consumption.

For the purposes of this paper it is important to pay attention to the position of government spending -- which is the last, "balancing" component on the expenditure side.

In the former socialist countries the macro-equilibrium was presented by a completely different equation:

$$GSP = G + C =$$
$$= [(ëI + I') + G'] + C, \text{ where}$$

- $GSP$ is *Gross Social Product*;
- $G$ is *all types of government consumption* (including *ëI - net investments, I' - depreciation* and *intermediate products, G' - other types of government consumption*);
- $C$ is *consumption of individuals*.

On the left side, the equations differ in qualitative terms because of the different concepts for productive and non-productive labor in the Marxian and in the neoclassical economic theory. According to the Marxian theory, only labor in the so-called "productive sphere" (agriculture, industry, construction, and "productive" services like transport, communications, and part of retail trade) contributes to a country's national income and social wealth. The quantitative consequence is that GSP does not count the contribution of the so-called "non-productive services" (education, health care, business services, recreation and sports, the non-productive part of retail trade).

On the right side, the equations reflect the difference between market and centrally planned economies. In the former socialist countries, based upon state ownership, government consumption included investments, and had a priority over individual consumption. An additional quantitative difference comes from the centrally planned economies' accounting system, which operates only with gross volumes, and therefore counts the intermediate goods more than once (increasing GSP in comparison to GNP).

At least two problems with long-lasting effects can be indicated. The first, easier to solve, is the artificial separation of net from gross investments and combination of depreciation with intermediate products consumption. The result is intensification of net investments as a source of growth and at the same time an ineffective use of depreciation funds (with a negative impact on natural resources and environment).

The second is with regard to the privileged position of government consumption. The monopolist power of the state gives almost unlimited opportunities for an expansionist growth policy.

To achieve a higher standard of living, the socialist society had to mobilize its productive resources and seek fast economic development. This strategy was distorted on ideological ground. Short-term high growth empirical data from the period of extensive industrialization were extrapolated for the future. Thus, the presumption for "ever-growing and cyclicly resistant rates of economic growth as inherent to socialism" was not only introduced in the economic theory, but endorsed as a positive (proven by facts) statement. Higher than socially justified rates of "rush" growth were accepted in the five-year plans, systematically destroying leading macroeconomic proportions (between investments and consumption, industry and agriculture, heavy and light industry, etc.). Indeed, an artificial macroeconomic equilibrium was constantly achieved through the redistributing mechanism of the budget.

The growing asymmetry between increasing investments and decreasing consumption finally leads to a point when no more resources can be moved from consumption to investment. In Bulgaria, in 1988 the rate of productive investments reached its upper limit. It pushed down investments and the rates of growth. Now the heritage of overinvestment in the past creates tremendous difficulties for potential restructuring, therefore imposing severe constraints on environmental policy-making.

One of the consequences of the investments-consumption asymmetry is the development of a specific psychology of underconsumption. In fact, the former socialist countries are now showing much stronger consumption-oriented patterns of behavior than the consumer society in the West. Two factors -- a long history of scarcity, and non-participation in investment decision-making -- diminish the entrepreneurial initiative. A strong desire to increase consumption drives income in consumer spending, and increases the investment difficulties of all former socialist countries.

As all reserves for growth have been used in the past, and as individual behavior favors consumption to investment, very little room exists for environmental incentives. Deep restructuring in ownership, incomes, and income distribution is required to liberate the state from the trap of being the main producer and the main consumer. It is highly improbable that the disequilibrium "investments-consumption" will be corrected in the short or medium term.

Nor can borrowing from the international markets be used to compensate for the internal investments shortage. Bulgaria has an $11 billion foreign debt (mostly owed to private financial institutions). As in the rest of the former socialist world, the expected environmental aid and credits are incompatible with the huge investment requirements.[23]

A realistic environmental policy approach needs to consider, first, the inevitability of a decrease in environmental spending, and, second, the necessity of integrating environmental concerns into the general economic restructuring incentives.

Not concentrating on environmental policy *per se*, but on efficiency-increasing industrial policy is the rational choice of the former socialist countries. Therefore, we need to shift from a macro- to microeconomic level, and analyze what are the options for pollution reduction at the firm level, and what obstacles prevent the polluters from realizing these options.

## Microeconomic Constraints

On a microeconomic level there are also barriers to environmental adjustment. The shortage economy left a disequilibrium between demand and supply, making the transition to a market economy inflationary, even without the pressure of grossly overvalued currencies. Classically, both demand and supply are inelastic, with considerable distance between them, forming the shortage area. Freeing prices, absolutely necessary for the transition, drastically reduces demand and only slightly increases supply.

Unfortunately, environmental resources pricing was not introduced in the very beginning of price liberalization; thus it is now viewed as a threat both for the consumers (trying to cut spending) and for the producers (trying to cut costs and to keep prices as low as possible). In an economy now careful to avoid price shocks while being extremely sensitive to inflation signals, environmental resources pricing is hardly acceptable and not clearly a positive step. An integrated approach (charges raise together with, instead of after, the major price shock) would have allowed switching over to near-full social costs pricing without the stubborn resistance by producers, consumers, and government officials now in action.

From a theoretical point of view, the lag in price adjustment of the environmental sector has two consequences. First, the environmentally related activities are deprived of the pillow of forced savings, which softens the first price jump and injects some money flows in the rest of the economy. Second, at a later stage the mass impoverishment drastically reduces both the ability and the willingness to pay (WTP) for environmental protection. If, as broadly accepted in the neoclassical environmental economics, the benefits from environmental protection are to be measured by WTP, virtually any policy step could prove economically inefficient.

Similarly other transition adjustments have had the same effect: first targeting the economic, and at a later stage, the environmental sector. For example, the withdrawal of state subsidies initially affected the raw materials producers, and then -- sewage plants and waste disposal firms, thus imposing barriers to price increases for waste water treatment and waste collection. Since almost 95 percent of the industries still belong to the state and the chances of quick privatization are marginal, this price pattern is likely to persist. It is advisable, however, to start considering and slowly introducing regulatory procedures and economic incentives for changing the behavior of the firms and the households.

## Environmental Policy in Crisis and Transition: Some Conclusions

How is it possible to promote the desired behavioral changes for protecting the environment in the wrongly-developed countries in Central and Eastern Europe? A potential answer is by building and implementing a modest, realistic environmental policy, integrated within the economic reform. Several steps in this direction can be suggested.

First, institutional changes need to be made to provide an operational system for environmental management and control. Retraining the regulatory staff (a process already started with international support) will provide the capacity to merge and analyze environmental and economic information, to deliver it to other decision-making bodies, and to use it in the decision-making process within the environmental sector. Major improvements can be achieved by decentralizing the environmental regulation, by more intensive cooperation with the key institutions of the transition process (the Ministry of Finances being the most important one), and by developing good monitoring and research units to back up the decision-making.

Second, it is important to clarify that making an implementable regulation is the most urgent need. The existing strict standards cannot serve as a basis for enforceable environmental policy, and therefore they need to be revised according to existing monitoring and financial capabilities. An evolution from lower initial standards, meeting the most urgent health protection needs and the capacity of the monitoring system, to stricter standards over a period of five to seven years is desirable.

Both suggestions require strengthening the regulatory mechanism. This seems contradictory to the logic of the deregulation reform. In fact, the command-and-control system of the past was so badly created, that the basic regulatory components (monitoring system, data bases, decision-making bodies, operational control units, etc.) need to be reestablished.

Third, the most powerful tool for reducing pollution is reducing the inefficiency of postsocialist economy, eliminating waste, and substituting highly polluting activities with cleaner materials and processes. In some cases simple changes in the production process can substantially decrease pollution, and improve efficiency, with only a marginal increase in costs. Therefore, key environmental policy instruments could be taxation (promoting pollution-free activities, clean products and introduction of best available technologies), export/import regulation, and subsidies reduction. If the key economic Ministries are active in supporting environmentally sound restructuring, the still minor role of the Ministry of Environment in decision-making can be tolerated.

Fourth, special attention has to be paid to the introduction and implementation of economic incentives. At the moment their "menu" is limited to fines, and factory close-downs. Pollution charges are also advanced (first for water pollution, and later for air pollution). The experience with imposing fines and closing-down polluting firms shows that affected industries are quick to mobilize support from municipal authorities and from the public (sensitive to raising unemployment) to resist and effectively stop or postpone sanctions. The suggested water pollution charges are being adjusted to the polluters "ability-to-pay", and designed in a way that makes them closer to fines than to internalizing pollution externalities or Pigovian taxes.[24] Without a strong public support the exclusive use of sanctions provide more incentives to illegal dumping than to technology improvements and pollution reduction. On the other hand, there is very limited understanding of how market-based instruments, like tradeable pollution permits or pollution taxes, can be developed and implemented.

Therefore, a successful promotion of economic incentives would first require more advanced "marketization" of the economy. In particular, property/use rights on the environmental resources should be clarified while the privatization process is underway. Massive state ownership in industries and non-defined rights and responsibilities for the state of the environment discourage policy incentives. Once a significant business community is established, however, it will be crucial to have a regulatory framework already in place, to transform pollution reduction and prevention in economic benefits by trading the "savings" or by the use of a refund scheme, as well as to dismay environmentally unsound economic behavior.

A major issue is the policy reaction to growing unemployment and "jobs versus the environment" trade-offs. Still very little is done to show that environmental regulation can create new industries, jobs and markets (for pollution prevention equipment, for substitutes to polluting products, for complementary goods). A well designed taxation and price policy can shift demand toward cleaner products, and stimulate their production -- for example a tax on leaded gasoline would not only encourage unleaded gasoline consumption, but also would create a market for catalytic convertors practically overnight.

Public support and public participation are the most important factors for successful transition. The environmental sector also relies on them. Education and technical assistance for compliance with new regulations should be provided free of charge. The new environmental policy should be designed in a simple and positive manner, and should be well elaborated in order to be understood and accepted. Writing it in a comprehendible way is an obligation of the new regulators in Central and Eastern Europe, for which help from the more experienced West is very welcome.

# Notes

1.  Bulgarian TV Program 1 Broadcast, March 24, 1992.

2.  See Bulgaria: *Crisis and Transition to a Market Economy*. A World Bank Country Study. Vol. II. Washington, D.C. 1991; *Godishnik za sastoianieto na prirodnata sreda na Republika Bulgaria - 1989 (State of Environment in Bulgaria Yearbook - 1989)*. Sofia, 1991; *Godishnik za sastoianieto na prirodnata sreda na Republika Bulgaria - 1990 (State of Environment in Bulgaria Yearbook -1990)*. Sofia, 1992; *Okolna sreda i razvitie na Republika Bulgaria (Environment and Development in*

*Bulgaria)*, A National Report.  Sofia, June 1991; Report No. 10142-
BUL of the World Bank (for official use only), November 26, 1991.

3.  The most typical example is the siting of Kremikovtsi steel mill.
    Despite scientific evidence, proving that the local iron ore reserves
    were of low quality (content about 30-32 percent iron ore per kg of
    extracted row material), the mill was built, now causing a high level of
    air pollution in the valley of Sofia.

4.  The scale of this problem is well illustrated in attempts to describe the
    environmental parameters of the Bulgarian economy in international
    comparison.  Key indicators like GDP, fixed capital, environmental
    equipment, etc., are quantified on expert judgment basis, and in the
    same time the level of pollution is calculated according to highly
    questionable information.  As a result, for example, Bulgaria ranged
    No. 1 in the world in $SO^2$ emissions per unit of GDP in a recent World
    Bank study (Report No. 10142-BUL).

5.  *Dargaven Vestnik na Tsarstvo Bulgaria (Bulgarian Kingdom State
    Official Journal)*, No. 59/1936.

6.  *Darjaven Vestnik na Narodna Republika Bulgaria (People's Republic of
    Bulgaria State Journal)*, No. 47/1967.

7.  *Dargaven Vestnik na Narodna Republika Bulgaria (People's Republic of
    Bulgaria State Official Journal)*, NNo. 19/1957;  51/1958;  9/1970;
    14/1973; 15/1982; 2/1987.

8.  *Godishnik za sastoianieto na prirodnata sreda na Republika Bulgaria -
    1989.  (State of Environment in Bulgaria Yearbook - 1989)*.  Sofia, 1991,
    p. 38.

9.  *Statistitcheski spravotchnik (Statistical Manual)*, Sofia, 1990, pp. 86, 87,
    99.

10. *Godishnik za sastoianieto na okolnata sreda na Republika Bulgaria
    (State of Environment of Republic of Bulgaria Yearbook)*, 1992, p. 130.

11. Interviews in the Ministry of Environment, March 1992.

12. *Dargaven Vestnik na Republika Bulgaria (Republic of Bulgaria State
    Official Journal)*, No. 86/1991.

13. By the time this paper was edited the Amendments to the Environmental Protection Act have not yet been passed through the Parliament. All comments on them are based upon officially circulated drafts, and press remarks.

14. In fact, the new water and air charges and fines are the old ones, multiplied by 10. Why this multiplier is selected is hardly to say. Presumably, it is to exceed the rate of inflation for a period of time.

15. What is questionable, however, is if the public wants to know - in other words, if enough is done to promote public interest. In two years all but one environmental periodicals in Bulgaria bankrupted. The last left is struggling to survive, meeting very little support from potential sponsors. At the same time the rest of the informational sector (TV, radio, press) is still not "environmentalized" enough.

16. See Gertcheva, Dafina and Dora Yordanova, *Reshaping Environmental Impact Assessment in Bulgaria*. Discussion draft paper, Sofia, 1992.

17. Earmarking becomes very popular in the environmental sector of the former socialist countries -- earmarked environmental fund already exists in Poland; other countries are in a process of creating it. The economic literature, however, is more cautious in respect to earmarking (see for example William McCleary, 1991. "The Earmarking of Government Revenues: A Review of Some World Bank Experience". *The World Bank Research Observer*, Vo. 6, No. 1 (January 1991), pp. 81-104.

18. The decentralization in Central and Eastern European context can also be dangerous, because the totalitarian period left limited managerial capacity at local level. Another reason to question its large scale implementation originates from the resource scarcity. There are fears that the decentralization of the environmental fund may lead to fragmentation of money and may damage large environmental projects. The new revision of EPA, initiated by the first non-communist government, among other tasks also tries to find the delicate balance between centralization and decentralization.

19. Proposed Amendments to the Environmental Protection Act, as formulated by the Environmental Committee of the National Assembly. The Ministry of Environment suggested even more centralized structure -- 70 percent in the National Fund versus 30 percent in the Municipal Funds.

20. A quantitative indicator is the share of "green" seats in the Parliament, which dropped from 9 percent after the first democratic elections to 4 percent after the second one.

21. For 1991 the lag between expectations and reality for all macroeconomic indicators was substantial -- annual decrease in GDP 20 percent (expected 11 percent), monthly inflation rate of 4-5 percent (expected 1-2 percent); unemployment rate - 12 percent (expected between 2 and 6 percent); budget deficit - 7.5 percent (expected 3.5 percent); exchange rate - 21/23 leva per dollar (expected 7/10 leva per dollar) - see *Konunkturni pokazateli (Current Economic Indicators)*, No. 1/1992. Agency for Economic Programming and Development, Sofia, January 1992.

22. See also Georgieva, K. *"Eastern Europe: The Collapse of the Totalitarian Socialism"*. Working Papers No. 18/1990, SSED, University of the South Pacific, Fiji.

23. As S. Kabala points out, "while the cost of clean-up is tallied in hundreds of billions of dollars, available foreign assistance is counted in hundreds of *millions* (Kabala, Stanley J. *"Environment and Development in the New Eastern Europe - Addressing the Environmental Legacy of Central Planning"*. In: *Occasional Paper - 3*, Geonomics Institute, Middlebury, Vermont, February 1992).

24. The suggested water pollution charges should be calculated per cubic meter of used water. This is a highly disputable solution, since (i) water tariffs are still below the full costs of water; and (ii) the charges are related neither to pollution abatement costs, nor to level of pollution. Further work on the charge system can be expected.

Chapter 5

# PUBLIC PARTICIPATION LAWS, REGULATIONS AND PRACTICES IN SEVEN COUNTRIES IN CENTRAL AND EASTERN EUROPE: AN ANALYSIS EMPHASIZING IMPACTS ON THE DEVELOPMENT DECISION-MAKING PROCESS

Stephen Stec

The environmental recovery of Central and Eastern Europe critically depends upon well-reasoned and fair decisions by authorities with respect to all activities affecting the environment, taking into account their short- and long-term impacts. Experience in the West has proven that the best way for decision-makers to avoid mistakes based on lack of information is for the public to actively participate in the decision-making process. It is equally important to those involved in the re-development of the countries in the region that the wishes and concerns of the public be anticipated and addressed at an early stage of the process in order to promote efficiencies, avoid delays and to increase certainties.

The right of the public to truly participate in decision-making must be guaranteed by law. Many of the states of Central and Eastern Europe and the former Soviet Union have some existing legal basis for public participation in decision-making, for example, the People's Councils. But in past practice, these institutions usually rubber-stamped decisions from above. They may provide the structure, however, for the future growth of public participation rights. In addition, lessons learned from Western cultures may be adaptable to these countries in transition.

*Public Participation Research Project*

In Spring 1992 representatives of the European Bank for Reconstruction and Development (EBRD or the Bank) and the American Bar Association's Central and East European Law Initiative (CEELI) met to discuss the possibility of conducting a joint project aimed at assessing and surveying public participation institutions (as established through laws, regulations and practices) in Central and Eastern Europe. The Bank was particularly interested in public participation issues relating to investment decision-making, with the long-term goal of promoting the exchange of information and helping Central and East European countries to improve public participation in such decision-making processes. CEELI was a natural partner for assisting the EBRD in achieving its goals, since CEELI, a law reform and information exchange assistance project, had placed experienced

A. Vari and P. Tamas (eds.), Environment and Democratic Transition, 88–119.

American lawyers with access to governments and the larger legal communities in each of the six countries (Albania, Bulgaria, the Czech and Slovak Federal Republic, Hungary, Poland and Romania) that are the subject of this paper. (CEELI now also has resident lawyers placed in each of the Baltic countries and several countries of the Commonwealth of Independent States [CIS] and is continually expanding.) Furthermore, CEELI's goal of improving public participation institutions and practices through law reform, though not specifically limited to considerations of the nature of investment decision-making, was consistent with EBRD's stated goals.

*Project Method*

After the two organizations agreed to work together, the project organizers developed a questionnaire aimed at gathering basic background information about laws, regulations and practices in the area of public participation. Several basic areas of inquiry were identified. They included constitutional and related provisions (such as statements of basic rights); environmental laws; specific laws concerning environmental impact assessment (EIA); the general administrative or public administration law; freedom of information and right-to-know laws; laws and practices pertaining to permitting of projects (e.g., land use and construction permitting on the state and local level); and non-legal public participation alternatives (including public demonstrations, use of the press, etc.). Wherever possible, the questionnaire included requests for the respondents to provide copies of applicable laws, regulations, guidelines, etc., in English if possible, or in the national language if no English translation were available.

Once the questionnaire was finalized, the organizers went about the task of identifying respondents in the six countries in which CEELI had resident liaisons. (The EBRD also used the questionnaire to gather information in other countries, including the Baltics, former Yugoslavia and some CIS countries.) The goal was to identify four respondents representing a cross section of viewpoints in each country to give as accurate and balanced a picture as possible. The profiles for the respondents were as follows:

1. One academic involved in studies or drafting of laws in this field.
2. One upper-level official in that ministry which is responsible for privatization or is, otherwise, most closely involved in public participation in the development process.
3. One mid-level or local administrator who makes decisions about development.
4. One representative of a group that has attempted to participate.

Identifying adequate respondents for each of these groups proved problematic. One limitation was language, since the language of the project was English. In some cases, CEELI liaisons had translation and interpretation resources available, but the extent of these resources varied greatly from country to country. Another limitation was expertise. In some cases, there simply was no person in the subject country who met a given respondent profile. A third limitation was time and money. Some potential respondents, although willing to assist, were severely limited in the time they could give. A few potential respondents refused to cooperate on a voluntary basis and would only consent to respond if paid. In such cases, alternative sources of information were found, generally with no loss of accuracy.

The responses to the questionnaire were analyzed and checked against the language of the subject country's laws and regulations. In addition to a formal review process involving the CEELI in-country liaisons, other opportunities for review by experts from the subject countries were taken as they presented themselves -- for example, through informal consultations during regional workshops and conferences. A "final" set of country reports was produced in July 1992 and was incorporated into a broader EBRD report that included similar information on additional countries, including republics that made up former Yugoslavia, the Baltics and some CIS countries.

### Project Follow-up

Given the rapid pace of change in this area, as well as in all areas that are at the basis of reform toward modern participatory democracies, information is continually being updated. Several of the subject countries are now in various stages of considering draft legislation that will drastically affect their public participation institutions, and coincidentally, thereto the contents of the country reports. Consequently, the project organizers recognize that the information contained herein, even if it is as accurate now as can be expected under uncertain and difficult circumstances, will very soon be obsolete. It is, therefore, the hope of the project organizers to continue the project with the continued support of the Bank, CEELI, and other interested organizations. The goal of the Public Participation Research Project is to use the knowledge gained from this and further surveys to promote the growth of legal and non-legal institutions through which the general public and, especially, potentially affected parties can provide information to and participate in the making of decisions affecting the environment, both in relation to development projects and in other areas as well.

# Albania

## Introduction

In Spring 1992, the Socialist Party (formerly the Communists) lost control of the government. The new government is just beginning to address its environmental policy, and their efforts are hampered by the absence of a tort system. Furthermore, meaningful public participation is included only in a few areas of the law. An international drafting effort, with the Committee for the Protection of the Environment (the Committee, under the Ministry of Sanitation, was created by decree in May 1991), is underway. Although it is premature to assume the outcome of the drafting process, a draft of the law has been analyzed and is discussed below.

## Constitution and Related Provisions

Article 16, No. 7491, of the August 1991 law "On the Main Constitutional Dispositions" stated the national policy that environmental protection constitutes a prerequisite for sustainable development of society and shall be a priority of national interest. Pursuant to this statement, the People's Assembly is charged with drafting an environmental protection law.

## Environmental Protection Law

### Existing law

Pending passage of the new law, there is no modern comprehensive environmental protection law in Albania. Prior to the recent changes, a number of decisions of the Council of Ministers concerned national parks, fishing, food safety and public health, including regulating exposure of workers to radioactive substances. A 1991 decree (No. 7451) established guidelines "For the Protection and Preservation of the Environment," amending the former major environmental law dating from 1973. Environmental authority is vested in the Committee, whose mandate is to oversee investments for the protection of the environment, to monitor enterprises which pollute (and to close them if necessary), to organize educational activities and to publish scientific articles regarding environmental protection. The Committee has an information branch to provide information to "foreign entities." Public participation is not provided for in any existing environmental law. One inspector for the Committee, however, has stated that theoretically it is the responsibility of the Committee to hear complaints about decisions affecting the environment.

The Ministry of Construction, responsible for parks, maintaining apartment buildings, and disposing of industrial waste may also be responsible for enforcing environmental laws and regulations. This Ministry has no enforcement staff or resources at present, however.

Under decree No. 5105, all foreign investors must obtain a permit from the Committee, but the public does not formally participate in the permit process.

### The proposed draft law

Under Article 29 of the draft law, control over (presumably meaning enforcement against) the sources and causes of pollution shall be exercised by the Minister of Health and Environmental Protection

at the request of natural and legal persons and citizens affected or that may be affected by environmental pollution and damage, as well as other organizations of an environmental character.

The Committee of Environmental Protection shall have the duty to organize the "propagandation" of education and participation of the population in environmental protection. Article 40, Paragraph 6.

A private cause of action for damages due to environmental pollution is established under Article 43 of the draft.

### *Environmental Impact Assessment*

### Existing law

There is no law concerning Environmental Impact Assessment.

### The proposed draft law

The current draft law contains the requirement that all activities of natural and legal persons in Albania shall be subject to EIA. The breadth of that statement is qualified in further provisions specifying which activities shall be subject to compulsory EIA. Article 11 of the draft provides:

Natural and legal concerned persons shall have the right to participate in the process of the consideration of results and environmental impact assessment.

They shall be informed by national or local mass media or other appropriate means about the procedures of environmental impact assessment, not later than one month prior to its start.

Further the draft places responsibility for informing the population about the local environmental situation and activities subject to EIA in the local authorities. Article 42, Paragraph 4.

*Right-to-Know and Freedom of Information*

### Existing law

Albania does not have any law requiring the government to inform the public about environmental data, or giving the public a right of access to government information. According to one government official, it is no longer the policy of the government to deliberately mislead the population about pollution, but at the moment the alternative has been to provide no information at all.

The Committee's Information Department could, in theory, provide the information it gives to foreign entities to the citizenry as well. One source within the Committee stated that it is the department's intention to provide such information, but that the public has never asked. Instead, the public may ask the Council of Ministers or another governmental agency.

### The proposed draft law

The draft law places responsibility for control of the "environmental situation" on the Minister of Health and Environmental Protection, who is obliged to gather information to form the basis for his decisions in fulfillment of his duties. Persons whose activities affect the environment are obliged to report environmental data to the Minister, as well as to the Committee of Environmental Protection, its regional branches and other authoritative bodies, according to rules defined by the Minister. Article 34 of the draft provides

The bodies referred to ...shall give publicity to the information, by mass media or by any other means, in a form accessible for the citizens, information which contains data on the change of environmental situation.

Trade secret protection is provided for as well. The next article elaborates:

The relevant state bodies, as well as the natural or legal persons immediately after causing pollution and damage to the environment, shall be compelled to inform the population about the occurred environmental adverse alterations [and] appropriate measures about the conduct of the citizens as concerns health protection and their security.

Failure to report information about adverse effects on the environment caused by a person's activities, failure to advise the public about actions to be taken by citizens to protect their health and safety, and failure to warn about the dangerousness of goods and services are violations punishable by administrative fines under proposed Article 45, Paragraphs 5-7.

A general statement protecting the consumer is found at Article 36:

The natural and legal persons shall inform their buyers or customers during the time of sale or performance of service, in writing or orally about these components of goods and services which are dangerous and about the possible adverse effects and impacts on environment and human health.

### Public Administration Law

Such laws do not provide mechanisms for public participation.

### Centralized and Local Permitting

Permits are issued by central government agencies according to zoning master plans. These plans do not address environmental issues. Permits for construction may require the approval of the Ministries of Construction, Economy or Agriculture, for example.

Local governments have virtually no power over permitting, although the Executive Committee of a district may be consulted.

The public has no right to information concerning permitting decisions, nor is there any right to challenge permits on environmental or other grounds.

### Public Actions

There are two major green factions, the Green Party (aligned with the ruling Democratic Party) and the Ecological Party (aligned with the Socialist Party). Individuals have given information to these parties about such issues as hunting licenses and activities of foreign oil companies, but it is not clear what steps these organizations are taking to influence the government.

According to one source within the emerging Green movement, there have been no significant public protests or actions yet, although public opinion is becoming increasingly sensitive to environmental issues. Another source, however, said that protests have taken the form of written submissions to government officials on topics such as hunting by foreigners and irresponsible fishing practices. One protest concerned citizens cutting rare trees in the Botanical Garden for housing purposes. In Elbasan, a particularly polluted area, the public has organized "green" associations to a comparatively great extent. Albanians have been watching with interest the developments in other East European countries.

# Bulgaria

## Introduction

Change in Bulgaria has been rather rapid. After initial elections in which the former communists retained power, there has been a recent change in that government has turned power over to the former opposition. The populace is comparatively aware of environmental and development issues due to the important role played by the green movement in the initial phase of transition. New laws on the environment and on public participation establish democratic institutions, and a law giving greater access to information is in the drafting stage.

## Constitution

The Constitution adopted in July 1991 sets forth general principles regarding the environment and provides a general guarantee of access to information by the public. Article 55 states:

Citizens shall have the right to a healthy and favorable environment corresponding to the established standards and norms. They shall protect the environment.

Article 41 (2) provides:

Citizens shall be entitled to obtain information from state bodies and agencies on any matter of legitimate interest to them which is not a state or official secret and does not affect the rights of others.

## Environmental Protection Law

Bulgaria has a new Environmental Protection Law, enacted in October 1991. Under sections 8 and 9, all persons have the right of access to information about the state of the environment, which includes:

data about the result of actions, causing or likely to cause pollution or damage to the environment, or to its components;... [and] data about activities and actions, undertaken with the purpose of protection and restoration of the environment.

The law further requires the government to gather and requires producers of goods and services to provide such environmental information. Apart from the more specific EIA requirement discussed below, environmental impact information about projects is also available upon request prior to issuing final operating permits. The right to disseminate such information may be limited by other provisions of law, such as those relating to trade secrets.

Further information disclosure requirements relate to the occurrence of accidental releases and the hazardousness of products entering commerce:

[Authorities, as well as producers of goods and services] shall inform the population without delay when pollution or damage of the environment occur, including natural disasters, industrial accidents and fires, and shall provide information about the changes in the environment that have taken place, the measures for their restriction and elimination and the requirements for the conduct of the citizens with a view to ensure their health and safety.... An authority or a person who considers that his request for information is unjustifiably rejected or unlawfully restricted, or that the obtained information is unreliable, shall have the right to request protection of his rights through administrative channels or through the court.

### Environmental Impact Assessment

Section 20 of the Environmental Protection Law provides that *all* activities may be subject to an EIA requirement. EIAs are obligatory for national and regional development programs, territory-structuring and urban-development plans, and their amendments; for projects for reconstruction and enlargement of existing enterprises included therein; and for specific types of projects enumerated and registered pursuant to an appendix to the law. Furthermore, in certain instances, local government bodies have discretion to order an EIA. The Bulgarian Parliament recently (in late Fall 1992) substantially amended the EIA provisions of the environmental law. The amendments added provisions giving a right to interested physical or legal entities to propose to competent authorities that an EIA be ordered. Also, certain facilities with large environmental impacts must perform an EIA every five years.

Under section 20:

All concerned physical and juridical persons shall have the right to participate in the consideration of the results of the environmental impact assessment....

Concerned persons shall be informed of an EIA procedure at least one month before it commences, through the mass media or in another appropriate way. The competent body shall make conclusions based on the EIA, taking into account the views of the persons participating in the procedure. Concerned parties may appeal the findings, within two weeks for local projects or within four weeks for projects of national significance. The

Minister of Health shall review a negative EIA finding, and may require an assessment of the state of health of affected persons as well. The proposed amendments would move these provisions to new sections 23a and 23b as discussed below.

The Minister for Environment, with approval of the Council of Ministers, is due to issue an EA Implementation Scheme. According to sources, competent bodies, including the Regional Inspectorates of the Ministry of the Environment and local governments, do not yet have sufficient experience working with the EIA requirements and procedures.

New sections 21-23b further spell out the obligations of the investor or proposer of the activity in undertaking the EIA process. New section 23a requires the competent authority to organize the discussion of the results of the EIA. The discussion shall include:

(1) The organs of the local administration, representatives of NGOs, the public and interested physical and legal entities.

(2) The persons under paragraph 1 must be informed by the investor or initiator of the activity by means of the mass media or other suitable way not later than one month before the discussion.

Under section 23b the competent authority must adopt its decision within three months after the holding of the hearing. Within 14 days after adopting its decision, the decision must be published in the same manner as above. Interested parties may appeal the decision to the regional court under normal laws of administrative procedure within 14 days of the publication of the decision in matters of local importance or within 30 days of publication in matters of national importance.

### Right-to-Know and Freedom of Information

An information law is currently in the draft stage. The current draft contains general statements giving the right of free access to information which is the subject of a person's legitimate interest. The law would place an obligation on government to disseminate any information it has that is of public interest.

### Public Administration Law

The 1991 Law on the Management and Administration of Municipalities provides for direct public participation in matters of local importance. It obliges local authorities to notify the public about their decisions. Meetings of municipal councils are advertized and open to the public. Local representatives are required to inform their individual constituencies about council decisions. An earlier law, not followed, requires "adequate time-limits" for introduction of every project to the public.

By way of example, in the municipality of Bourgas, local environmental matters are considered and the laws enforced by the Supreme Ecological Council, together with the Mayor. A sub-department on the "ecological program" prepares plans and programs for the ecological development of the municipality, which is established through discussions within the councils which are advertised and open to the public.

On the local level, some municipalities are revising their local laws and ordinances to reflect changes in the national law. For example, Bourgas is drafting new ordinances to "reflect the democratic changes" and to reflect the requirements of the Environmental Protection Law. The new ordinances will specifically allow the public and NGOs to participate in local government proceeding, and will provide for the creation of public "green patrols."

Some attempts have been made to use referendum rights under the Law for Polling the People for environmental purposes, but they have not been successful due to the vagueness of the law.

In addition, there are normal rights of appeal from administrative decisions, including any decision of the Mayor of a municipality.

Although the Bulgarian Parliament is under a Constitutional deadline of July 17, 1992, to revise the Bulgarian judicial structure, no action has been taken yet. The matter is reportedly at the top of the Parliament's agenda for 1993. The new law will result in a completely new administrative procedure with as-yet unknown public participation provisions.

### Centralized and Local Permitting

Issuance of permits depends upon the type of development project being considered. Construction projects are generally covered by the Law on Territorial and Administrative Planning. The Council of Ministers has established a committee to approve large projects. A smaller committee within the Ministry of Building and Housing approves smaller projects. A representative from the Ministry of the Environment sits on these committees. Projects solely within a municipality are approved by the municipal councils, with a representative of the regional commission participating.

Typically, development project proposals must first be approved by the municipal Department of Urban Planning or other relevant department, and then by the municipal Department of the Environment. There are no specific public participation procedures applicable to these proceedings, although in many cases there are open hearings held and the press reports on them.

A 1988 decree on the public assessment of the programs, plans and projects for construction and urbanization provides another means of public participation. Although there is no requirement that the public be notified of specific permit applications, there is an opportunity for public participation in the overall development plan for the municipality through the annual publication of the Urban Plan. A one-month comment period is provided after publishing the plan. Permit applications consistent with the plan need not be published, but permit applications that would effect a change in the plan must be advertised. The public has a right to challenge permits issued if they are not consistent with the Urban Plan.

The new law on Economic Activity of Foreign Persons and on Protection of Foreign Investment requires that permits for certain foreign investments must be issued by a committee established by the Council of Ministers. There are no specific provisions for public participation or consideration of environmental matters in this law, but the law generally provides that Bulgarian law shall apply to foreign investments unless otherwise specified.

## Public Actions

The recent change in government resulted in a green party becoming an important player in national politics.

There is a fairly well-developed recent history of public actions. Protests against the construction of the Balene nuclear power station included a strike action in early 1990. After eight months of similar protests, the Council of Ministers suspended the project. Civil disobedience resulted in the suspension of lead smelting activities and the implementation of health protection schemes at a lead and zinc plant in 1990. A plan to divert the flow of the Rila River to Sofia (Rila-Mesta Hydro) and construction of the Cherni Osam Dam were also suspended after protests.

# The Czech Republic and Slovakia

## Introduction

The Czech and Slovak Federal Republic separated into two nations at the beginning of 1993. The two new nations have started off, at least for now, by continuing the environmental laws in force prior to the breakup. But, even in its previous form, two distinct and somewhat autonomous republics, peculiar problems in administration of public participation in the development process were apparent. The former federal framework and structure applies, but it has remained and still remains within the competencies of the new nations to give substance to the framework laws. Consequently, the state of practice was uneven, as between the Czech Republic and the Slovak Republic during the period of union; and, subsequently, the separation of the two republics will result in two distinctly different environmental legal frameworks. Recent laws have been enacted which seek to guarantee public participation in important governmental decisions, including those relating to development.

### Constitution and Related Legislation

Article 35 of the Basic Rights and Freedoms (added to the Constitution in 1990) proclaims that everyone has a right to accurate and full information on the state of the environment and natural resources. Public participation is a component of the Constitutional Act introducing the Basic Rights and Freedoms.

The Act on Petition Right and the Act on Assembling Right contain instructions for public participation concerning their respective subjects.

### Environmental Protection Laws

On the Federal level, the CSFR enacted laws, including the General Law of the Environment, the Clean Air Act, the Law of Waste, and the Revised Building Code. The Czech Republic also has adopted the Law of Environmental Impact Assessment and Law of Nature Protection.

The General Law of the Environment (Article 14), adopted in December 1991, provides:

Everyone has the right to true and accurate information about the state and development of the environment, the causes and consequences of that state, activities which are being prepared and which could change the environment, as well as to information about measures taken by the authorities responsible for environmental protection in order to prevent or remedy environmental damage. A special regulation may stipulate cases in which such information can be restricted or withheld.

Procedures and regulations governing this law have not yet been promulgated. The practical effect is to put all enterprises on notice that the public has some right to information and that steps should be taken to prepare for the time when the law will be implemented.

Furthermore, the law obliges citizens to participate. Article 19 provides:

Everyone who learns about a threat to the environment or about environmental damage is obliged to take such measures that are within his or her powers to eliminate the threat or minimize its consequences and to report the facts without delay to the state administrative authorities.

Enforcement is on the republic level, through the Czech Ministry for the Environment and the Slovak Commission for the Environment.

## Environmental Impact Assessment

An EIA process is required for a wide range of proposed activities, including land development, use of natural resources, construction and changes to industrial activity.  A governmental review committee on the federal level considers EIAs.  According to Article 25 of the General Law of the Environment:

The extent of the environmental impact assessments shall be discussed by the assessing authorities with the relevant state administrative authorities, with the communities whose territory is to be affected by the impact of the plan and with the general public. The completed environmental assessment shall be subject to similar discussions.

## Right-to-Know and Freedom of Information

There are no specific freedom of information laws.

## Public Administration Law

Appeals to administrative courts may be taken under current law only when procedural rights have been violated, and not in cases where the authority did not correctly assess facts in the matter.  In addition, an "extraordinary" appeal may be taken by the Prokurator.

## Centralized and Local Permitting

Under current rules of administrative procedure as applied to local land use planning and development projects, the applicable construction office announces the opening of the permitting procedure directly to participants in the process, overseeing negotiations among interested parties.  In some instances, a broader public notice may occur.  From the time of notification, interested parties shall have a minimum of seven days within which to comment.  Current rules of administrative procedure apply to this process.

Environmental administrative decision-making is concentrated in district authorities.  These decisions are made on the district environmental department level in the Czech Republic and subdistrict environmental office level in the Slovak Republic.  In the Czech Republic, district decisions can be appealed to the Czech Ministry of the Environment.  Slovak subdistrict environmental office decisions can be appealed first to the district environmental office and then to the Slovak Commission for the Environment. Local and municipal governments are limited to an advisory and consultative capacity.

Under the 1976 Act No. 50 on Land Use Planning and Construction Rules, persons whose property rights may be directly affected by a land use planning decision have the right to participate in such decision-making through a kind of EIA procedure. In such proceedings, municipalities are also interested parties, in their own capacity and as representatives of the broader public, where the proposed decision relates to a large construction project, such as a dam, nuclear plant, etc.

*Public Actions*

A number of Green organizations exist, both local and international. Public demonstrations are not uncommon, but require a "parade permit." In 1990 Greenpeace draped a great protest banner from the cooling tower of the nuclear power plant under construction at Temelin. In 1991 Eurochain and others organized several actions to protest the completion and putting into operation of the canal and Danube dam at Gabcikovo, in the Slovak Republic. In heavily polluted Northern Bohemia, grassroots organizations successfully urged the government to raise prices of coal and energy and to establish a Czech Government Commissioner to address the unique problems in this area.

# Hungary

*Introduction*

It could be expected that Hungary, with a relatively strong and complex economy, and a great deal of foreign interest in investment opportunities, would serve as a model for the rest of Central and Eastern Europe in matters relating to environmental protection and public participation. An international environmental law-drafting effort, through an independent working group of Hungarian legal experts, was launched in 1991. The product was supposed to be a model for the region. That draft, commendable but not without its critics, continues to be refined. Meanwhile, the Ministry of Environment and Regional Planning has renewed its own drafting efforts, based in part on the independent group's findings. A draft should be presented to Parliament early in 1993. After a great deal of behind-the-scenes wrangling, the current Ministry draft contains substantial public participation provisions. It is expected, however, that the draft will undergo substantial change before being enacted.

## Constitution

Chapter I, Article 18 of the Constitution of the Republic of Hungary (1990) states: "The Republic of Hungary shall recognize and enforce the right of all to a healthy environment."
Chapter VII, Article 35 (1) states:

The Government shall:
a/ protect the constitutional order, protect and ensure the rights of citizens;
b/ provide for the enforcement of the laws;
c/ direct the work of the Ministries and other bodies directly subordinate to it, and coordinate their activities.

Article 36 of the same chapter states:

In the course of discharging its functions, the Government shall cooperate with the interested social organizations.

Chapter IX, Article 42 states in part:

The exercise of the powers of local self-government shall be an independent, democratic administration of the local public affairs concerning the community of the electors and the exercise of local public authority in the interest of the population.

## Environmental Protection Law

The effort to draft a new comprehensive environmental protection law for Hungary, which would serve as a model for the rest of the countries in Central and Eastern Europe, has proven to be a difficult process. It is now an optimistic prediction that a new comprehensive law will be passed sometime during 1993.

### The proposed draft law

The public participation provisions, which form an integral part of the early drafts that have been presented and considered, are particularly controversial. The drafting committee, with advice from Western sources, presented clear, well-developed procedures for meaningful public participation. But the administrative procedures law currently lacks the sophistication to give a context within which meaningful public participation can take place. Many persons recognize the need for a new administrative procedure act, yet they do not believe the environmental law is the proper means to change administrative procedure. The current Ministry draft would create certain administrative procedures and rights applicable only to environmental matters.

For example, one version of the draft includes provisions (22(a) f.), requiring the government to publish proposed decrees, regulations and standards for public comment, prior to their adoption.

Under subsection 134, Paragraph 2 of the current draft:

The participation of citizens in the formulation of decisions about the environment and protection of natural resources -- including governmental organizations and local government activities in the protection of the environment -- is secured by law.

Section 135 would provide that citizens may establish associations for representing their interests in environmental protection, these associations having, *inter alia*, the following rights: [to]

(b)participate in procedures defining sites for those developments and capital investments concerning essential interest of their members;

(c)participate in environmental protection official  procedures concerning their area of operation; and

(d)make use of public communication media to demonstrate environmental damages and dangerous situations.

Furthermore, proposed subsection 136 would cast these associations in an extraordinary role.

(1)  The association has the right to employ all legal instruments in the interest of environmental protection, to require governmental associations and local governments falling within their competence to take the necessary measures.

(2)  In case of default by a competent state authority or local government to take the necessary measures, the association has the right to initiate a legal proceeding against the user of the environment.

Hungary is also considering creating the position of Environmental Ombudsman, who would protect citizen interests in environmental protection, analogous to other Parliamentary Commissioners of Citizens Rights.  There are currently several alternative viewpoints as to the ombudsman's responsibilities.

## The existing law

The environmental protection law currently in force (adopted in 1976) is typical of such laws under the former regimes, giving lip-service to strict environmental standards, while providing no mechanism for administration or enforcement. There are no provisions in this law for public participation in decisions relating to environmental impacts, environmental protection or other decisions that might be made during the development process. Enforcement of the current law is undertaken by the Regional Environmental Inspectorates, 12 in number, which also handle permitting of facilities, in coordination with the regional Public Hygiene authorities and the regional Water Management authorities. There has been talk about coordinating permitting (including the EIA component) under the single authority of the REIs, although this will be debated.

Appeals from decisions of the REIs are heard by the Chief Environmental Inspectorate. Further recourse can be had in the courts. Citizens do have the right to urge action by the REIs or to complain about REI decisions. The Director of the REI should answer complaints within 30 days, after which the citizen can complain to the CEI. If no satisfaction is forthcoming, the citizen can then write a letter to the Minister of Environment. According to Ministry officials, this process is actually used somewhat frequently, but rarely results in any real change, as the Ministry takes a somewhat paternalistic approach to such questions.

### Environmental Impact Assessment

## The existing law

EIA is not mandatory, and permits may be issued even if an EIA is lacking.

EIAs are, however, often done, especially when driven by foreign investment decisions. According to one Ministry official, EIAs are being carried out because locals and western investors are requiring it through other means (see below). The State Property Agency, in charge of privatizing individual businesses, has no strict guidelines concerning whether to order an EIA; and, in fact, the current process is one in which the individual officers who negotiate privatizations have broad discretion about the terms of the agreement.

The Ministry released guidelines on EIA in April 1991 -- Technical Instructions of the Ministry of Environment and Regional Policy on EIA. Although not a legal document, it extensively considers public participation. For example, it recommends proper presentation to interested economic and public organizations of information, data and all EIA documentation. The guidelines give instructions as to procedures and feedback mechanisms as well.

## The proposed draft law

The draft would initiate full EIA requirements for certain projects, under subsection 88 *et seq.*. Although there is no express public participation component in the EIA section in the draft, procedures for EIAs, pursuant to proposed subsection 58 *et seq.*, include public notice, the opportunity for comment, and public participation in hearings. Participation in hearings would require a threshold determination that the requesting party is "interested," but, under other provisions of the draft as it currently stands, meeting the requirements for such status would not be difficult. Section 83 would provide that appeals may be taken from EIA decisions, including the right to appeal from decisions denying the right to participate and the right to appeal a decision where interested parties could not participate because of the negligent failure of the authorities to comply with publication and information requirements.

### Right-to-Know and Freedom of Information

In October 1992 the Parliament passed a law (Act LXIII of 1992) on data protection that included some provisions on publicity of public-interest data. That law declared that the "most important" data in the public interest shall be made available to the public. Members of the public are granted a right of appeal to the courts in circumstances where officials fail to observe the law. The law contains an exemption for state and service secrets, however. It remains to be seen whether such a vague and general law will be of any use to public participation in environmental considerations.

Data on pollution and health matters are now public and are required to be disclosed under the 1991 act on the Public Health Service. This has been interpreted not to apply, however, to data on individual facilities. Rather, the National Public Health Service is required to publish periodically a report on the general conditions relating to health and the environment, with overall levels of pollution for areas within Hungary. The current draft environmental law still contains an EIA component that will include right-to-know provisions.

Another problem, however, is whether good data will exist. Apart from resource limitations, which are to be expected, under current law the REIs do not have inspection rights that would allow them to enter privatized premises. They may only monitor water discharges, for example, at the property line. Also, there is no requirement that industrial accidents be reported, although such circumstances normally go into calculation of fines.

Practice can often result in disclosure of information not required to be disclosed by law. For example, early in 1992 a small explosion occurred in a chemical plant in Budapest. The Lord Mayor then published a list of the biggest polluters in a newspaper of large circulation. Similarly, it is fairly common for citizens to ask local authorities for hearings on certain matters. Though this may not be a formalized process, it can result in the disclosure of useful information.

The proposed draft environmental law would require all data (except those protected by trade secret limitations) acquired during environmental monitoring to be accessible to the public (Subsection 137).

## Public Administration Law

The current law on public administration, amended in 1981, contains a general statement that:

the administrative procedure is based on the effective cooperation of the public authority, of the participating parties and of any other organs and persons participating in the procedure.

Participating parties are guaranteed certain procedural rights to ensure that their views are at least considered by relevant public authorities; but, in reality, this does not guarantee meaningful public participation. According to practice, "participating parties" have traditionally consisted of the central or local authorities and certain state-organized interest groups, such as government agency representatives, workers' associations, and scientific institutes. Thus, this process more closely resembles an internal governmental consultative process than it does a public forum. Moreover, it is virtually impossible to get an outside organization added to the "flow chart." There is no obligation in law or practice to give public notice of most administrative procedures, or to invite interested parties to testify or make requests to participate.

For example, a local self-government recently considered transformation of a vacant commercial property to a trucking transfer station. Residents of the immediate vicinity had been urging the local authorities to convert the property into a park. When news of the proposal to make the trucking depot leaked out, the local residents protested. They were told by the local authorities that they were not interested parties, even though the new function of the property would result in increased traffic congestion, noise and pollution. This case has not yet been resolved.

Given the current position of local and central authorities as to who is a participating party, it is apparent that some provisions of the public administration law are virtually meaningless. A person who proves that information is necessary to protect a legal right may examine government files and make copies, but the authorities can refuse to disclose such information, 1) where they determine that the requesting party doesn't have a recognizable legal right; or 2) in the interests of official secrecy. There are no enforceable guidelines at present that would limit the authorities in declaring something an official secret.

### Local Self-Government Law

Under Act LXV of 1990 on Local Self-Governments the local self-governments within the Republic of Hungary were granted autonomy over a wide range of matters affecting the local populace. Among the tasks of the municipal self-governments (numbering in the thousands) are, according to Chapter II, Article 8 (1):

local development, resettlement, the protection of the built and natural environment, housing management, water management... [cooperation] in solving problems of employment;... facilitation of the establishment of the communal conditions of a healthy way of life.

Under paragraph 2, the self-governments have discretion to decide what tasks they will execute, to what extent and in what manner, according to the local requirements and financial limitations.

The local representative body shall hold public meetings, although upon justification it can hold closed sessions (Article 12[3]). Moreover, under Article 13,

the representative body shall hold public hearings announced in advance at least once a year, where the citizens and the representatives of local organizations can ask questions and make proposals on matters of public interest.

A significant provision is Article 18(2):

the representative body shall determine the order of the fora (village or town policy forum, town district meeting, village meeting, etc.) the purpose of which is to inform the citizens and social organizations directly and to involve them in the preparation of important decisions. The representative body shall be informed of the standpoint taken by these meetings and of the minority opinions expressed.

Thus, the potential exists, if local self-governments do their duty, for there to be a meaningful forum with good public participation on issues of local importance, including (under Article 8) issues relating to local development, the natural and man-made environment, and employment.

Finally, under Article 45 *et seq.*, the public has rights pertaining to referendum and initiative. Under the proposed draft environmental law

(Subsection 131), the public would have specific referendum rights pertaining to the siting of public interest capital investments (such as solid waste landfills, waste-water treatment works, incinerators, etc.).

## Centralized and Local Permitting

The centralized permitting process currently involves all of the authorities mentioned above, plus others such as the Nature Protection Directorates, the Road Transport Directorates, the Regional Land Offices, labor health authorities, and in special cases, geological survey (mining), civil protection agencies (explosives, nuclear), the Ministry of Defense and the Ministry of Interior (boilers and heating systems).

Local governments have authority over projects through two main mechanisms -- land use permitting and construction permitting. Although there is no requirement, other than what is contained in the Local Self-Government Law, for public participation in local permitting decisions, grass-roots political pressure can result in locals using their permitting authority to influence development. For example, the construction of a battery-recycling plant has been delayed, and is not yet complete, because of local procedures. A nuclear power plant project, which involves the proposed disposal of low- and mid-level radioactive wastes has been effectively halted because the local government refuses to issue the land use permit.

## Public Actions

Public awareness seems to have ebbed a little after rising during the early days of the new regime due to the almost daily attention on the Danube Dam issue. A number of Green groups flourish, however, and sponsor activities such as bicycle rides and tree plantings, many of which were celebrated on Earth Day 1992.

Public protests contributed to the controversy over the international agreement between Hungary and CSFR over the construction of hydroelectric dams on the Danube River. At present, the Hungarian government has disavowed the agreement, but construction continues on the Slovakian side of the border. The Hungarian government has decided to pursue the matter before the International Court of Justice. The result of public protests against a battery recycling facility and a radioactive waste dumping site have been mentioned above.

Another example occurred with regard to a closed heavy metal plant. Lead contamination was discovered in the soil in the vicinity of the plant. At the urging of the green movement, public hearings were held. The mayor, acknowledging that the land was unmarketable, compensated the residents by reducing their land tax to zero.

# Poland

## *Introduction*

Poland has a relatively long history of modern environmental law, which is a matter of pride for many Poles in the field in spite of the disastrous state of the environment in many parts of Poland. Although commendable for its idealism, if nothing else, the result of these early efforts is that the need for a new, modern comprehensive environmental law has not been universally recognized. Instead, improvements have been made piecemeal through amendments to the original law and through scattered and diverse decrees. This creates difficulties for the "regulated community" since there is no single, clearly defined set of guidelines in one environmental law. Moreover, the laws and regulations that exist are of the socialist-era type -- that is, ideologically they pay lip service to protecting the environment, but in reality, they provide only a sparse practical mechanism for achieving ecological ends and assuring responsible governmental action.

## *Constitution*

There are no constitutional provisions guaranteeing public access to information or public participation in governmental decisions.

## *Environmental Protection Law*

In 1991 the Parliament adopted the National Environmental Policy as proposed by the Ministry of Environmental Protection, Natural Resources and Forestry. In pertinent part, the statement of principles reads as follows:

Principle of active participation of citizens and public organizations, expressed by various forms of public inspection of environmental protection, the universal right to advance claims aimed at abandonment or limitation of actions against the environment, and the universal right of access to information about the state of the environment and the means of its protection...

Principle of local self-government participation in securing environmental protection, understood as a gradual process in the local administrative entities due to their strengthened position and elevation of expertise.

The Act of January 31, 1980, on Protection and Development of the Environment (Article 99) provides:

Agencies of villages or municipal districts, workers self-management, trade unions, and other civic organizations interested in environmental protection in connection with their activity may undertake activities improving protection of the environment and the right of citizens to utilize its benefits.

## Among the specific activities enumerated are the following:

4/ presenting proposals towards preserving and improving the environment in investment and mining activities;
5/ cooperating with state-owned enterprises in preparing and implementing projects and plans for improving the environment;
6/ increasing the public influence towards more effective environmental protection activities of companies;...

It is unclear whether "other civic organizations" could be interpreted to include private NGOs.

### Environmental Impact Assessment

Under Article 70 of the same law, environmental impact assessments are required for a great number of projects. Article 100 covers public participation through civil suits (see below) and includes provisions relating to participation of the public in EIA proceedings. The latter provisions apply to projects with "great" environmental impact.

2/ Before issuing the investment decision on a particular investment which has great environmental impact, the state government administration agency which makes the decision as to the placement and the construction site of the investment, shall inform appropriate civic organizations about the proposal. Within the specified time, not exceeding 30 days, these organizations may present their remarks and objections to the proposed investment.

3/ The appropriate state government agency shall consider the remarks and objections and shall inform civic organizations which delivered them.

EIA requirements were further defined in the Executive Order of the Minister of Environmental Protection on investments particularly harmful to the environment and public health and conditions required for evaluation of environmental impact of investments and buildings, prepared by experts, Monitor Polski (1990). Further regulations about EIAs are being prepared. The 1992 Law on Nature Protection also contains EIA-related requirements, as does the 1984 Land Use Planning Act, discussed under "Centralized and Local Permitting."

Direct public participation (as opposed to participation through a civic organization) is provided for under current law only in cases posing extreme environmental and health hazards. In other cases, individual members of the public may petition to be admitted to participate as interested parties.

The Minister has established an independent, 75-member Commission for Environmental Impact Assessment. The Commission's meetings are open to regional and local authorities, local environmental groups and the press.

For projects posing extreme hazards, the Commission selects a group of 15-25 from among its members to conduct a review of the EIA in consultation with experts. The Commission's opinion is proclaimed at a public hearing, then transmitted to the Minister for final decision. The decision is published in the Commission's Bulletin, available to the public. EIA's for other projects are reviewed on the regional government level.

## Right-to-Know and Freedom of Information

Under Article 100 of the Environmental Act and Articles 38 and 42 of the Planning Act, "civic organizations" have rights to information about investments. This could be interpreted to grant the public, through NGOs, for example, the right to know about environmental aspects of investment projects.

It has been practice recently to disseminate, through the press, national and regional lists of major polluters. These lists are compiled by the State Inspectorate for Environmental Protection and the regional government (voivodship) offices.

## Public Administration Law

Under Article 100 of the Environmental Law, civic organizations may seek court review of administrative inaction in the face of environmentally dangerous conditions. In the first place, civic organizations may complain to the proper state administrative agency and may appeal administrative decisions to the High Administrative Court, pursuant to the Administrative Code. The Administrative Code provides that all organizations may participate as a party in any proceeding. Interested parties may also petition the court directly to order cessation of the harmful activity and to further order restitution or award damages.

## Centralized and Local Permitting

The Act of July 12, 1984, on Land [urban and rural] Planning, as amended by the Act of May 17, 1990, on Division of Duties and Competences in Certain Acts Between Local Self-Government Agencies and State Government Administration Agencies and on Amendments to Certain Acts (the Land Use

Planning Act), governs centralized and local permitting. Under the law, local governments have primary authority to issue land use permits for construction of greenfield industrial facilities or expansion of existing facilities. An EIA procedure is required for projects with potential impacts harmful to health and the environment.

Articles 15 and 28 elaborate the procedure for creating local development plans, including the right of "interested agencies and organization units of state government administration, other organizations and private individuals" to submit remarks and proposals on the means of development of a particular area. This information shall be disseminated by announcement in the local press, advertisement or in any way customarily used in a particular locality. Local self-governments and some other agencies have to be informed in writing.

Article 30 further provides for a 21-day notice and public comment period for the draft plan. If a specific submission is not adopted, the local government must inform the commenter in writing of the reasons therefore. It has been reported that the law allows for public participation in individual investment decisions as well.

The Regulation of the Council of Ministers on the Classification of Investments and the Subject, Rules and Procedures for Investment Location Decisions (1990) classifies various proposed development projects as national, regional or local. According to the classification, national, regional or local governmental bodies have virtually autonomous power over issuing permits for investment projects. Local self-governments have substantial role in decisions on local investment. Voivodships (regional administration agencies of central government) have a substantial role in decisions on regional investments, pursuant to the Planning Act. And central government ministries have substantial role in decisions regarding national investments, also approving all regional land development plans. The Central Planning Office is responsible for national and regional development plans and the Minister of Land Development and Construction is responsible for local plans.

Various state agencies have competencies within the investment process and may stop the investment for noncompliance -- e.g., State Inspectorate for Environmental Protection, State Sanitary Inspectorate.

*Public Actions*

Successful public actions against construction of a nuclear power plant in Zarnowiec took place in 1989-90. The public succeeded in shutting down the Skawina Aluminum Works. Public protests in 1991 against the building of a dam in Czorsztyn were unsuccessful, and the dam is currently under construction.

# Romania

## Introduction

The free flow of information and the right of the public to take part in decisions affecting their health and welfare basically do not exist in Romania at present. An environmental law is currently in the draft stage and would have some as-yet unclear public participation provisions. A legacy of the Ceauceascu regime is the theoretical framework for localized decisions through municipal councils, which, although idealogically tainted in the past, could perhaps be improved to provide for meaningful public participation.

## Constitution

Article 31 provides that the public has a right of access to information of a public interest and imposes upon the government the obligation to ensure that the citizenry is informed about matters of public affairs and personal interest. This obligation also extends to the public and private mass media.

## Environmental Protection Law

Under the current Romanian environmental protection law (1973), including Decision No. 264/1991, concerning the setting up of the Ministry of the Environment and its main tasks, there are no provisions for public participation in permitting decisions, nor are there provisions giving the public any right to information about specific facilities. The current law does contain a general statement declaring that the public has a right to be informed about environmental issues, but this is not interpreted to give access to information about individual permitting decisions or projects. A new draft law is being prepared, which at present contains provisions for public participation and right to information. The government had hoped to enact the new environmental law in 1992. Drafting is the responsibility of the General Division for Strategy, Legislation and International Projects within the Ministry.

Chief among the specific provisions of the draft having to do with public participation in the development process are provisions guaranteeing the right of interested persons to information and consultation concerning decisions for land reclamation projects, and siting potential industrial developments with negative environment impacts. Further, the draft would grant a right to all citizens, directly or through an association, to challenge authorities regarding actions with negative environmental impacts.

*Environmental Impact Assessment*

Under Decision No. 97/1991, concerning the elaboration and approval of the technic-scientific documentation and the opening of financing for new investment objectives, and under Decision No. 264/1991, some documentation is required, but according to a high-level ministry official, real EIA requirements and procedures will be developed through specific regulations after the adoption of the new law on the environment. It is expected that local residents, NGOs, recreational associations and trade unions will have the right to participate in the EIA process.

*Right-to-Know and Freedom of Information*

There are no right-to-know or freedom of information laws that would require public disclosure of facts or issues relating to development decisions.

*Public Administration Law*

Romania has an ombudsman system which allows for the hearing of individual complaints on local matters, in addition to the normal appeals process from administrative decisions. The public also has the right to make law through referendum.

The 1991 Law on Public Administration requires local councils to act in order to restore and protect the environment, parks and nature reserves, and to conserve and protect historical and architectural monuments.

*Centralized and Local Permitting*

Local and municipal governments have substantial authority over issuing permits for smaller investments. On the local level, authorities are reputed to have wide discretion with regard to the settlement of various matters which have potential impacts on development decisions. Larger investment projects (above 500 million lei, or involving more than 50 hectares or arable land or 100 hectares or forest) require central government approval. Institutions which may have consultative or approval status include the Local Agencies for Environmental Survey and Protection, District Centers for Prophylactic Medicine, District Inspectorates for Silviculture, District Bodies of the Autonomous Regions, and regional water authorities. According to one source, the local community has the right to information and consultation concerning decisions on land planning and locating potential industrial development. Another source stated that the public is informed "incidentally about issuing permits on new investments, taking into account the capacity of the new objective and the possible impact on the environment."

### Public Actions

The green movement is in its infancy. There are no reports of significant protest actions, except for general references to public demonstrations and hunger strikes.

## Conclusions, Trends and Anticipated Developments

As is apparent from the foregoing, many and varied forces are at work in just the seven subject countries -- from the extreme example of Albania, virtually cut off from the rest of the world until a time so recent it can still be measured in months, taking the first tentative steps toward instituting a legal framework for environmental concerns, to the relatively sophisticated situations in Poland, the Czech Republic and Hungary. One thing these last three have in common is a developed technical and scientific infrastructure. Yet they must be distinguished from each other in their solutions to current environmental legal problems. Poland's gradualist approach contrasts starkly with the Czech Republic's activism in the last few years. In Hungary gradualism has been taken to new extremes and can best be characterized as intransigence. Slovakia has the benefit of inheritance of the first basic steps towards institutionalization of environmental protection, but political pressures bearing on a new nation trying to find its way could derail further progress. Romania appears mired and directionless. The remaining country considered, Bulgaria, has adopted an impressively simple, yet strong, environmental protection act, and has already amended it based upon considerations demonstrating actual experience from implementation and practice.

Other countries not covered in the survey, but located in the larger region of Central and Eastern Europe, face their own difficulties in integrating public participation components into their environmental laws. The states resulting from the breakup of the Soviet Union are preoccupied with economic transformation. Those new states involved in the war in the former Yugoslavia must secure the peace before they can hope to solve environmental problems. Even so, in almost all the countries in the region, there are some initiatives to modify existing laws or to enact new ones. For example, new comprehensive laws are in the drafting stage in Slovenia and Croatia, and Lithuania is considering a new law on nature protection.

Although efforts to improve legal solutions to environmental problems continue, the initial euphoria that followed in the wake of 1989 has worn off, and attention has turned away from the environment to other matters, including the strengthening of social safety nets to catch those displaced by economic transformation.     With the slower pace of environmental law reform, public participation mechanisms are slower to be developed as well.

In Albania the international effort, led by the World Bank, will assist the new government in formulating its environmental policy.  The new law which will result from this effort should include a public participation component similar to the provisions analyzed in the draft law.  It is clear, however, that environmental protection is not a priority of the government, nor is it yet a matter of strong public feeling.

The Parliament will consider a draft law on local governments, which promises to de-centralize decision-making to some extent.    Local governments currently have virtually no authority.

In Bulgaria an information law is in the draft stage.  The current draft contains general statements giving the right of free access to information which is the subject of a person's legitimate interest.  The law would place an obligation on government to disseminate any information it has that is of public interest.

There is a proposal to enlarge the information required to be reported and disseminated under section 11 of the Environmental Law to include assessments of the current state of "components" of the environment, and data about protection and restoration activities.

Laws are currently being drafted on forests, hunting, fishing, waste treatment, the marine environment, and protected territories and objects (including national historical treasures).   Specific environmental laws on water, air and soil are also in the drafting stage.  Revisions will be made to the law on noise pollution.  A law on underground resources is virtually complete, and a law on coastal protection is under consideration.

Slovakia is working on a new constitution.

The new Czech and Slovak national governments are in the process of fulfilling the requirements of the General Law of the Environment, with regard to specific matters such as environmental impact assessments.  It is unclear what effect the recent elections and the separation of the two republics will have on efforts to implement the environmental law.

Prior to the elections, amendments were being planned to the EIA law in the Czech Republic, and a law of EIA was being drafted in the Slovak Republic.

Because of the delay in enacting Hungary's comprehensive environmental protection law, the Ministry of Environment and Regional Planning has decided to pursue the course of promulgating EIA requirements through ministerial decree.  This decree is intended to be a temporary stop-gap measure    reflecting    current    legal    requirements,    and    the    Ministry acknowledges that it will become obsolete upon the passage of the new comprehensive law.

The following laws are currently in drafting or on the agenda for future consideration in Hungary:

- EIA law (with right-to-know provisions)
- Comprehensive Environmental Protection law
- Construction law
- Physical Planning and Zoning law
- Nature Protection law
- Preservation of Historical Monuments law
- Animal Protection Law

Laws currently in the pipeline in Poland include the following:

- Draft environmental protection law (intended to replace and unify existing law -- includes proposal for citizens' suits to enforce environmental laws)
- Draft Building Code (in Parliament)
- Draft land development planning act
- Draft water law
- Draft waste management act

The Romanian draft environmental law has been under consideration for some time. It has undergone many revisions. After elections in late September 1992, the new government underwent a reorganization and eliminated the Ministry of Environment as an independent ministry, dividing its former competencies among several other ministries. This reshuffling obviously works against the process of adopting a comprehensive law.

In all of the countries surveyed, the next few years will be critical, both from the point of view of development of law reform and in the larger historical context. If the economic transformations take hold, then attention will again focus on environmental and other law reform matters. If, however, political backlashes against dislocations rise to prominence, the economic debate will continue to be a long and drawn-out affair. Besides economic pressures, changes in the global and regional power structure have created vacuums that allow nationalism to rise and ethnic interests to re-emerge. War is raging in former Yugoslavia and in many spots in the former Soviet Union; the possibility of expanded conflict cannot be ignored.

Yet apart from anxiety, the rapid pace of unpredictable change in Central and Eastern Europe has also created an atmosphere of possibility. As creaky old institutional structures tumble or are dismantled, a new generation with new ideas is building new structures, sometimes from the ground up. The environment, decentralization, responsibility of government and public participation are components of the new agenda for change for many of the active visionaries in the region. They face entrenched corruption, apathy and social insecurity, but they do so with hope and determination, able to draw upon Western experiences, while still seeking solutions consistent with the unique histories and legal cultures of Central and Eastern Europe.

Chapter 6

# ENVIRONMENTAL INITIATIVES IN RUSSIA: EAST-WEST COMPARISONS

Oleg Yanitsky

This paper focuses on a wide variety of public participation processes, both in terms of subject (improvement of the living environment, the fight against pollution, protection of historical and cultural monuments) and goals (protection, renewal, reconstruction, innovation, and social experiments).

We are interested in comparing the dynamics of goals and forms of public participation in Russia to those in Western European countries. We are especially interested in the process of public participation which began to develop rapidly in Russian cities in the second half of the 1980s.

The research reported here focuses on a number of critical questions: What is the correlation between the Western and Eastern, particularly Russian, pattern of development of public participation? Are the evolving processes in Russia following the Western ones, and, if so, do they pass through the same stages and same forms? If not, what are the differences between these two types of processes? In their everyday activities, do Russian public groups consciously reproduce Western patterns, or is the activity of Russian environmental groups and movements the result of sharply changing national contexts? Can the dynamic of goals and forms of public participation in the West serve as a model for the East? Finally, the most important problem is whether concepts of social (environmental) movements developed by Western sociology are universal, and to what extent are these concepts applicable to the East European context?

## The Issue of Context in East-West Comparative Analysis

Several researchers (Castells, 1983; Pickvance, 1985, 1986a; Nelissen, 1991; Hackney, 1991; Krantz 1991; Deelstra, 1991), interpret a context as a set of permanent conditions under which initiatives and movements emerge and develop. Other authors (Touraine, 1984; Hegedus, 1989; Arato & Cohen, 1984) emphasize that new social movements, including ecological ones, influence the creation of a new type of civil society which has new interrelations with the state, different from the previous ones. In other words, in both the West and in the East (although in different ways) not only the context, but also the characteristics of the social subjects, i.e., individuals, groups and organizations acting in a society, are changing. Therefore, it is necessary to define the concept of context.

A. Vari and P. Tamas (eds.), Environment and Democratic Transition, 120–145.

A context is an environment for activity by a social subject which provides the subject with some resources or, on the contrary, demands from it some expenses. The activity resources of any social subject can be ranked according to the degree of control of the individual over them. At one extreme are internal resources of this subject which, if necessary, can be used automatically or with minimal expense. At the other extreme of the dichotomy is resources as fixed conditions (resources-preconditions) which the social subject has to take into consideration in his or her activity, to adapt to them, but which he or she cannot quickly alter (Yanitsky, 1986). At the same time, context can be represented as a system of catalysts and blockades; that is, conditions not directly influencing the subject but stimulating or limiting his or her activity.

These general and constant conditions for all social subjects are described as a context. The third feature of a context is its integrated impact upon the subject; that is, a context is considered as unified environmental conditions. The poverty or wealth of society, its stagnation, crisis or progress, peaceful or bellicose character, calm or tense political situation -- all of these circumstances designate different aspects of context. The type of society and public opinion are characteristic forms of context. Context can be global, regional or local; and the character of the impact which a context makes upon a subject depends on the type and goals of activity of the latter.

In comparative research, a context must be clearly analyzed and described. For example, if we take into consideration such an element of political context as a party system, then not only the presence or absence of a multiparty system is important, but also the significance of the fact of party membership in a given society, and the attitude of public opinion to a party as an instrument of political will.

Comparing the results of research into initiatives and movements in the West and in the East, we propose the following typology of contexts:

*Context I* is a historical-cultural context, a stable system of determined relations between state, civil society and population, and also of the cultural norms regulating them.

*Context II* is a macro-social context, i.e., it is a context that has both stable and innovative elements. This might be called the "context of a broad historical development" (Nelissen, 1991) or the context of a transitional period. For the West, this is a transition from industrial to postindustrial society, and, for the East, the transition from a totalitarian society to a democratic one.

*Context III* is a situational one. This context is an immediate (economic, social or some other) environment, in which initiatives and movements emerge and on which resource maintenance and development are directly dependent. The urban environment is an example for Context III.

The meaning of each of these elements of context is also very important. For example, multi-party systems in West and East European countries are not the same. Nonparticipants who have grown up in the conditions of market economies create a quite different socio-cultural background for the initiatives and movements under consideration, as compared with those who have grown up under centralized economies.

Now let us successively compare these three types of contexts under the conditions in the East and the West.

### Context I

The historical-cultural context of movements in the West and East is sharply different. In the first case, conditions are bourgeois-democratic, and, in the second, authoritarism and totalitarianism are prevailing. Despite repeated attempts by enlightened Russians to introduce elements of European civilization to Russia, beginning with Peter the Great, the divergence between European and Russian civilization is growing. Underlying the former are the interests of the individual, whereas the latter is founded upon the interests of the state.

Today's Western civilization is a liberal civil society based on private property, market economy, and civil rights and liberties institutionalized in constitutions. Political representation and expression of public opinion are characteristic of Western civilization as is a developed pattern of voluntary associations. The attitude of compromise operative in this society and numerous mediators between government and society are important for the emergence and maintenance of social movements. On the other hand, Western democracy is a socially and politically highly structured society.

Until now, Eastern civilization has seen the state dominate all spheres of social life; that is, an almost complete absence of civil society. Institutionalized public space and civil liberties have been practically absent. The degree of social and political structuralization has been rather low.

Differences in basic cultural norms are of major importance for understanding social movements in both types of society. Western society, despite the obvious struggle between different economic and political forces, is cemented by a deep consensus based on the civil rights institutionalized in constitutions and formal law, and also by political pluralism, decision-making systems allowing feedback, and by tolerance of different opinions. Such cultural norms as individual initiative and responsibility, self-reliance, and self-discipline consolidate Western society. Individual deviations, as well as evolutionary changes, are also seen as civilized norms.

In Eastern civilization, social life is cemented by state sanctions, the main goal of which is neither production nor innovation, but a distribution of resources. Executive behavior, equalization of distribution, and absolute subordination of individuals to the great mass (tyranny of the majority) are the norms which integrate Eastern society.   Norms appear such as "collectivism" (principles of depersonalized decision-making), and "centralism" (subordination to decisions adopted by the supreme power), resulting in the suppression of individuality, initiative, and responsibility.

In the West, social movements from the 17th to the 19th centuries created pluralist civil society and formal democracy.  New social movements strive for the creation of new solidarities and social space, and of new democratic forms (Arato and Cohen, 1984).  In the East, new social movements have no roots in the civilization.  Historically, at best, a network of small islands of civil society was created by enthusiasts, but, as a rule, a series of revolutions, forcible transformations, and wars ruined such islands of normal civil life.

## Context II

The question under review is that of the context in which the environmental movements have developed over the last 20 or 30 years.  At first this context seems to have many similar features in both the East and the West: a search for a way out from economic and social crises; a growing recognition of ecological crises; and a strengthening of interaction between countries.  A warming of the political climate in Europe has created a general attitude conducive to mutual assistance, cooperation, and partnership.  State and interpersonal contacts have been sharply intensified.

On closer examination considerable differences are seen in Context-II between the situations in East and West.  Although Western theorists distinguish between the situations of the 1970s and of the 1980s in Europe (Touraine, 1988; Hegedus, 1989), that was a context of continuity; innovations in the environmental sphere emerged and developed in the framework of a particular society.  In the East, Context II is divided into two periods:  the "context of stagnation" (the 1970s and the first half of the 1980s) and the "context of rapid changes" (the second half of the 1980s and the beginning of the 1990s).

The differences in Context II become much more considerable if the vector of the ongoing changes is examined.  In Western society, it is clearly a transition to postindustrial society.  In the East, changes are more complex.  For example, in Russia, a search and real changes are proceeding in at least three directions:  a rebirth of Russian national values and social structures (traditional peasant community, monarchy); assimilation of models of Western industrial civilization (private property, the market economy); and a search for the third way, consisting of technocratic illusions and noncritical perception of the short-term renaissance of civil society in the USSR in the 1920s.

The successive comparison of elements of the new environmental paradigm, as it has been formed in Eastern and Western public thought (Catton & Dunlap, 1978, 1980; Milbrath, 1984; Yanitsky, 1984), together with analysis of the disposition of the social forces which support or reject environmental values, (Yanitsky, 1991) suggests that Context II differences are much deeper. In the West, the vanguard (environmental reformer) takes political issue with the rearguard (traditional material wealth advocates, i.e., capitalists) (Milbrath, 1984). In Russia this vanguard is opposed to advocates of the administrative-command system, i.e., to defenders of the existing centralized economy.

The value contexts of a transitional period in the East and West are greatly different. In the West, notions of postmaterialist values and, in particular, of a new environmental paradigm, formulated by science nearly two decades ago have now generally penetrated mass consciousness. In the East, postmaterialist values are cultivated only by extremely limited circles of the public. For example, the bulk of the Russian population still adheres to the values of equalizing distribution and of executive authority, and these values are implemented through archaic forms of social action. Until now, the slogan of the Western ecological movement, "Think Global -- Act Locally," has been an exception in the political life of the USSR.

Deep differences in the management of the economy are also evident. In contrast to Western decentralized market economies, in Russia, the administrative-command system continues to dominate the economic sphere. No legal basis yet exists for economic decentralization, nor do necessary technological or human resources. The forms of decentralization of production that have emerged in Russia during recent years do not serve the purpose of raising its effectiveness and of encouraging initiative. Such decentralization is either a concealed means of preserving party-state property, of legalizing the shadow economy, or of achieving the private enrichment of the existing administrative apparatus.

Finally, while the West is looking for ways of creating a new type of civil society based on solidarity, legality, plurality, and openness (Arato & Cohen, 1985) or a creative society (Hegedus, 1989) or a "society endowed with historicity" (Touraine, 1988), Russia has only begun to fight for the creation of elements of civil society, for the very possibility of the autonomy of private life, for elementary guarantees of individual rights and liberties, for legality, for developing public opinion as a real social force, and for the mass media as an independent power.

Whereas the main innovation in the West has been the creation of a pattern of postindustrial society, the transformation of science and technology into a powerful source of social innovations, and the dissemination of postmaterialist values, in Russia the main innovation has been the realization that innovations are possible. Nevertheless, the degree of uncertainty in the economic and political situation is so great that a return to the norms and structures of totalitarian civilization cannot be excluded.

## Context III

Pickvance (1985, 1986) advanced five contextual factors that affect urban movements: rapid urbanization; state action (intervention in consumption, the character of responses to protest); political context (the ability of formal political institutions to express urban conflicts; cultural understanding about the scope of urban politics); and the development of the middle class and general socio-economic conditions (the general disposition to political activism prevailing at a given place and time). He argued that context affects these movements differently and stated that the relations between contexts and movements must be analyzed not from a static, but from a dynamic point of view.

At the same time, it must be kept in mind that Pickvance's approach was based mainly on Western experience. In the East, for example, there is no middle class as this concept is understood in Western sociology. So keeping in mind the specificity of the task, once again consider some questions of urban context.

Evidently, in both parts of Europe, Context III has much in common. This is predetermined by the common functional structure of modern cities and by the questions at issue -- pollution, transportation, overcrowding, ineffective self-government and the alienation of residents from their immediate environment. Hence, a common view of public participation as a medicine for improving the condition of the city organism emerges. At last situational context had much in common in that it mainly depended on local conditions, for example, the growth or decline of industry and the state of old city centers.

Differences between the East and the West are even larger. While the economic situation in West European cities has been more or less stable during the last five to eight years, in the East, and, in particular, in Russia, a complete collapse is observed. Up to the mid-1980's the cities' budgets, although meagre, were nevertheless regularly maintained by the state. The pressure of local party organs on enterprise administrators also helped in maintaining the city infrastructure.

From 1989, when the central government made decisions aimed at developing the self-sufficiency of enterprises, cities in general were deprived of financial support. This economic separation and the egoism of enterprises was strengthened by the weakness of newly-elected local authorities. After the democratic elections in 1990, a long-term conflict over economic and political power began between regional, city, and district authorities. In the absence of conditions of a market economy this has led only to the growth of inflation and a shadow economy. Financial sources which are routine for Western cities (property, land-rent, turnover, and other taxes) are absent in Russian cities.

Differences in the immediate political context also exist. Whereas in the West, a growing perception that political parties had been exhausted as mouthpieces for residents' needs could be seen (Touraine, 1988; Arato & Cohen, 1984), the processes of democratization in the East European countries have been accompanied by the emergence of numerous mini-parties. But the majority of these parties are either only mouthpieces for their leaders' political ambitions, or express the belief of ordinary people in representative democracy under conditions which they never experienced.

In the former Soviet republics the majority of these mini-parties are of the charismatic or populist type. The number of their members is defined by the popularity of their leaders. Besides, ninety percent of these parties are located in the capitals of the republics. Whereas enlightened capitals (Moscow and Leningrad) are relatively liberal, in the provinces, local municipal and party apparatuses are the leading antagonists of environmental initiatives. Whereas in the West, politicians and intellectuals are experimenting in order to prepare "a new phase in the local decision-making process" (Nelissen, 1991), in Russia and in the other former Soviet republics, the overall political context may be defined as a hard struggle.

Finally, attention should be paid to the socio-psychological aspects of the situational context. The responsibility and self-sufficiency which we consider as fundamental civilized traits and which have been inherent in Western citizens for centuries have always been combined with painstaking work at the grassroots level. State interference in private life and the alienation of residents from local decision-making acted as stimuli for grassroots initiatives. In Russia during the last decade, we have witnessed three different socio-psychological stages among urban residents. At the very beginning of Perestroika there was apathy and skepticism about any possible changes. Then came a psychological rise conditioned by the emergence of a new belief in the possibility of quick and painless positive changes. Now we are again witnessing the beginning of a new, much deeper psychological decline into pessimism and escapism. Nevertheless, the idea of the state being responsible for individuals' destinies was widespread in mass consciousness. A lot of ordinary people did not share in the compulsion to become active and to take initiatives, but held the position of detached onlookers.

## Participation and Resource Mobilization

Nowadays, resource mobilization theory is the dominant analytical tool in the analysis of social movements. Usually the central problem has been to explain individual participation in social movements. Such theories as those of mass society and collective behavior stressed the sudden increase in individual grievances generated by the structural strains which emerge under conditions of rapid economic and political change. Recent resource mobilization theory stresses the importance of such structural factors as availability of resources, organizational support, and political opportunities for movement emergence and maintenance, and emphasizes the rational motives of participation (Zald & McCarthy, 1979, 1987; Klandermans, 1984).

This section attempts to:

1) Reveal and compare the general resource structure of environmental movements in West and East European cities;
2) Determine the _differentia_ _specifica_ of environmental movement resource mobilization; and
3) Develop some new approaches to the mobilization problem.

### General Resource Structure

General resources of environmental movements include the following:

* Material resources,
* Information resources,
* Organizational resources,
* Political resources,
* Professional resources,
* Moral resources, and
* Temporal resources

Material resources are, first and foremost, money and the possibility of utilizing the existing urban infrastructure (housing, public buildings, utilities, transportation systems, and recreational spaces). Information resources include information about the concrete situation (in the city or region) as well as the possibility of utilizing the mass media for mobilizing other resources and for the organization of activities both of a creative and an emergency character.

The primary organizational resource is the existence and flexibility of supporting organizations and their mobility and adaptability to local conditions. The existence of external organizational support (from other groups or movements), as well as the possibility of relying on local organizations and of mobilizing informal connections is also implied.

The political resources of any social subject consist of access to power structures and of the possibility of participating in decision-making (through setting up pressure groups or by other means); political support of other external and internal groups and organizations; the state of public opinion; and the level of political activity of the local population (meetings, gatherings). This also includes methods and tools enabling an activist to neutralize opponents or even convert them into sympathizers. Finally, this includes political know-how, i.e., a set of tactics and programs making it possible to transform the demands of the urban population into political and other documents (drafts, resolutions, electoral platforms).

Professional resources consist primarily of the knowledge of prospects and scenarios for the development of situations and events in a city. A level of existing experience and knowledge of urban residents which is available for action in the given situation is also implied. Organizational knowledge and skills constitute an extremely important resource, especially in emergency situations. In mobilization situations, the availability of rapid access to scientific expertise (of the city, country, world), i.e., access to concepts, models, and data, is also an important professional resource.

Moral resources comprise first and foremost the values and norms of urban residents, their psychological readiness for mobilization, and the degree of development of their environmental thinking and ecological consciousness. Here, purely human qualities are important: the way people will act -- whether in line with their knowledge, convictions, and sense of duty, or in accordance with instructions and orders imposed from outside. The degree of dedication to goals is of considerable significance for the assessment of a groups' potential.

Finally, a special place belongs to temporal resources. This concept concerns the rate and timing of mobilization actions, and the ability to effectively utilize time. This ability depends on values and convictions, on personal knowledge and skills, the degree of personal self-organization and, of course, on the degree of resistance in the social environment.

## Resource Mobilization in Environmental Movements

Resource mobilization theory was mainly constructed on data relating to political protest movements, civil rights and alternative union movements, the peace, antinuclear, and anti-abortion movements (Zald & McCarthy, 1979, 1987; Klandermans, 1984). The specific character of resources and resource mobilization in environmental movements is defined by the objectives of these movements. Therefore, although individual initiatives and campaigns may be short-term and goal-oriented, the major aim of environmental movements is endless -- improving the living environment. So the resource mobilization of these long-term movements will be different from the patterns found within an initiative aimed at, for example, the removal of a military base.

In environmental, and especially ecological, movements, scientific knowledge plays a key role as a major resource and as a stimulus for the mobilization of other resources (an example might be the awareness of a particular environmental threat). "Ecological awareness... acted as a stimulus on the neighborhood committees to take action to improve the urban environment" (Bagnasco & Bonnes, 1991). Although, both in the West and the East, these movements have been partially transformed into political parties, they are mass movements because they are carriers of the needs and aspirations of the population of a given area. At best, a set of urban grassroots and environmental movements make up a united front of city dwellers.

In contrast to civil rights, peace, and union movements, whose demands are implemented via political actions, environmental movements require residents' direct participation in a variety of tasks, which, in turn, requires knowledge, skills, and know-how. As a rule, ecological and environmental movements are headed by professionals (intellectuals), but a lot of routine work is performed by ordinary people. Therefore, in these movements, concepts and models are as important resource as living experience and the desire to be informed about local issues.

In the West, resource mobilization takes place in a pluralist, democratic, and market society which has accumulated a lot of resources; and the main issue for the actor is to prove to society that he or she really needs this individual resource and is able to use it effectively. In Russia and some other Eastern societies, mobilization has, for years, meant top-down instructions implemented by, and in favor of, a state-party system. An endless set of top-down mobilizing organizations was the unavoidable condition of everyday life for millions of people. Under these conditions, mobilization was part and parcel of the usual system of compulsion applied to residents. They had no free choice between mobilization alternatives, and the results of a given mobilization were usually unknown. Even in the case of Chernobyl, a detailed accounting of the total sum of personal and organizational donations and of expenditures has still not been submitted to the public.

In Eastern societies, state and large branch organizations act as opponents of movements. In transitional industrial societies, such as Russia, the state becomes the core of overall domination. This confirms Touraine's statement that this "new role... has less to do with integration than it did in the past, and more to do with domination; it has become an instrument of power rather than of order; a mobilizer of resources, a manipulator of privileges, feelings, and political support" (Touraine, 1976). In the East, emerging social movements often have to operate in a situation of resourcelessness and, therefore, have to produce their own resources for their maintenance and survival. The major components of such resource self-production are moral motivation, self-exploitation, and informal networks of like-minded persons.

*Theoretical Considerations*

In the following, some theoretical points related to resource mobilization theory are summarized:

1. Resource mobilization is not a single act, but a process.
2. The notion of resource mobilization is no more than a conceptional frame for a set of processes which are highly dependent on the type of initiative or movement (protest, struggle, involvement, or decision making).
3. A distinction has to be made between the resource mobilization of the external environment and that of the social subject itself. The proportion depends on the type of the initiative and movement.
4. Resource mobilization is only one element of the process of emergence and maintenance of a movement. Movement emergence and maintenance imply also the accumulation, distribution, and the use of its results by the society.

# Emergence of Environmental Movements

The traditional approach to the analysis of social movements in Western sociology involves their division into two stages (emergence and maintenance/change) and two levels (micro- and macro-analysis) (McAdam, McCarthy & Zald, 1988). When it is applied to specific East European data, however, great difficulties immediately arise because, as was just shown, the context for which this approach was devised is quite different. In the East European case, at least three stages of emergence can be distinguished: 1) civil initiatives in an authoritarian society, 2) civil initiatives and movements in the initial period of democratization, and 3) social movements when essential socioeconomic and political transformations are already beginning.

*Civil Initiatives in an Authoritarian Society*

Contrary to the West, where there has existed profound division between public and private life and state and civil society, in Eastern Europe, the life of the individual has been strongly regulated by the state and its corresponding ideology. The tradition of local initiative and group activity was disrupted some 40 years ago in Eastern Europe and up to 60 years ago in the USSR.

In Eastern Europe, a qualitatively lower value was attached to the environment than in the Western democracies. This is not just a question of the population's far more modest material demands or the restricted opportunities for any environmental initiative. The wars, waves of repression, forcible resettlement, mass "voluntary" recruitment of citizens organized by the Party and other organizations (in the great construction sites of Communism and the opening up of the virgin lands), and forced rates of industrialization and urbanization were also responsible, as was the constant pressure exerted on the popular consciousness by the stereotypes of the levelling and collectivist ideology.

Nevertheless, environmental movements began to appear in the USSR before Perestroika. As we have just considered the problem of context in detail, we shall only briefly list the circumstances under which these movements emerged:

A centrally managed economy in which all productive assets belonged to the state; an incomplete form of industrialization and urbanization where living standards were low (by comparison with developed Western countries) and basic requirements (in accommodation, nutrition, and medical services) systematically went unsatisfied; the total lack or chronic underdevelopment of the elements of civil society; and an urban environment totally created by the state.

Politically the hegemony of the administrative-command system continues. For decades any local initiatives were assessed, first and foremost, in terms of their compatibility with the dominant ideology. Accordingly, all nongovernmental and "voluntary" associations were, in practice, part of the administrative-command system and were effectively under the supervision of the corresponding official bodies. Until 1990 there was no law concerning nongovernmental bodies; their activities were controlled by regulations adopted in the Stalinist period. There was no political machinery whatsoever for implementing grassroots initiatives. Public participation was widely proclaimed but there was, in fact, only a limited involvement of individual population groups in the activities of the party and administrative system. Urban politics in the Western sense did not exist.

For decades the social and functional structure of society was hierarchical and centralized. The Party-state apparatus, ministries and other central bodies were the key subjects, with the local Party bureaucracy, in shaping the urban environment. The informal group structure of society was very underdeveloped, existing, as it were, in the pores of the administrative-command system. Certain structures familiar in the West were absent: for example, local communities, municipal, and voluntary social services. A unified (Soviet) way of life and official collectivism (mass activities) were encouraged and local life had no autonomy whatsoever. In short, the socio-organizational structure familiar to the Western city was lacking.

In the first half of the 1980s, the majority of urban residents in Russia were tired and apathetic, and had exhausted their moral and physical resources; no one believed in the possibility of changes for the better. A 1984 survey in a major industrial center in the Russian Federation found that 75 percent of respondents did not expect any changes in their lifetime and did not feel that they, themselves, had the strength to implement them (Abankina, 1986). The fear of external (war) and internal (repetition of 1930s) threats persevered. Consumerist and egalitarian aspirations were strengthened due to the growth in the number of uprooted people (migrants, military personnel, those with nomadic occupations). There was a decline of interest in work. Various forms of escape were on the increase (retreat into private life, migration in search of high wages or some promised land, alcoholism, drug abuse, and other forms of delinquent behavior). The system of reciprocal services and other types of informal labor continued to develop. The first pre-Perestroika initiatives arose in an atmosphere of value vacuum: faith in the declared values of socialism was already practically exhausted, but a new value system had not yet been established. For the intelligentsia and other representatives of those middle-ranking social strata that formed the potential base for civil initiatives and local movements, this vacuum created a high level of psychological tension.

## Civil Initiatives and Movements in the Initial Period of Democratization

At the beginning of Perestroika, Russia did not possess the preconditions for the development of initiatives and movements that are typical of the majority of Western democracies. The context in which environmental movements emerged was qualitatively different from the West.

What then did Western and Eastern societies have in common? At the macro level, one could mention the penetration of the state into private life; the emergence of such new values as a desire for community life, self-actualization and personal satisfaction; ecological awareness, and a certain experience of self-organization (McAdam, 1988).

Here also, however, was a significant difference. If Western theorists, for example, mentioned the penetration of the state into private life, this meant, as everyone understood, penetration into those areas that had formerly been personal and private. In Russia almost all private life was supervised in one way or another by the state. Furthermore, ecological awareness, almost never resulted in a well-organized urban population in Russia. Finally, the level of prior self-organization is only a common factor in the sense already indicated, that small groups already existed in the pores of totalitarian society.

Without any doubt, the formation of small, informal groups with strong leaders is extremely important. We agree with the assertion that "informal groups or associational networks... are expected to serve as the basic building blocks of social movements." Indeed, they create the "cell structure of collective action" (McAdam et al., 1988). We also agree that "micromobilization contexts can be thought of as a dense network of intermediate-level groups and informal associations" (McAdam et al., 1988).

Five to ten years ago, however, East European environmental, ecological, and other initiatives emerged in conditions of the severest infrastructure deficit, especially in the Soviet Union. Education indisputably played a stimulating role; but, in the USSR, a number of initiatives arose because students and adults were dissatisfied with the quality of teaching and education. For example, voluntary student patrols for protecting the environment (part of the ecological movement which had already began in the 1970s) passed on to participants a great deal of practical know-how, filling a gap in their professional training.

Interestingly, environmental catastrophes did not play a major role in the emergence of movements in the USSR. Soviet people had lived through so many horrors and upheavals (war, famine, repression, forced migration) that a constant psychological readiness and preparedness for possible catastrophes, both personal and social, became part of their mass consciousness. Even a man-made catastrophe such as Chernobyl did not give birth to a serious anti-nuclear movement in the USSR; only five years later did the victims of Chernobyl begin to organize. As far as migration to the cities is concerned, these migrants were the most passive group in the urban population because they were totally dependent on their urban employers.

In general, Western models for the emergence of social movements are too calm and rational for the situation. In Western society, the emergence of such movements is a norm of political life and culture. The era of movements in Soviet society, however, began within the framework of the declining, but still extremely powerful, administrative-command system.

### Civil Initiatives and Movements During the Period of Democratic Transition

The key element for East European countries today is the relation between the sociopolitical situation at the local and national (ethnic) levels. In less than a decade, the countries of the region have passed through at least three stages. In the first, were attempts to start local initiatives, mainly those concerned with culture and ecology, when all of public life was still highly centralized. At the second stage, those same initiatives began to develop: Perestroika was underway but Glasnost and democratization had only been asserted from the top and had no economic or political underpinning at the local level. "Everything is permitted, but one can't do a thing" sums up the context of action for local initiatives at this stage. The next stage has been the emergence and very rapid development of social movements as major economic and political changes take place. In other words, the development of these movements takes place in a transitional period which is distinguished by an extreme and universal instability in the economic and political situation and a bitter struggle between the administrative-command system and the nascent cells of civil society.

Under such rapidly changing circumstances, it is natural that local initiatives, and the ecological and environmental movements that depend upon them, will also evolve rapidly and make changes in their goals. At the first stage, these were the semi-legal cells of civil society which had to fight off the administrative-command system on all sides. No matter how radical the thinking of their leaders might be, the activities of such initiative groups were generally adapted to the dominant sociopolitical system.

When, at the second stage, independent activity was permitted from above, these initiatives tried to legalize their activities and put their programs for regeneration and renewal of the environment into practice. But they came up against the unbending opposition of the Party-state apparatus. This led to a sharp politicization of all initiatives and accelerated their coalescence into social movements, unions, and blocks in order to put pressure on the central political authorities. As a result, for a certain period, the specifically environmental aims of the activities of these groups and movements receded into the background. Some environmentalists stood as candidates in the parliamentary and municipal elections; others canvassed for them; and yet others served as consultants and experts in drawing up electoral platforms and programs.

Finally, at the third stage, Green parties, or movements very like parties, emerged and their representatives were elected to the reconstituted republican parliaments and, in particular, to the local Soviets.

The most favorable context for environmental action is when the new local authorities are based on, or attract, the already existing neighborhood committees and other cells of civil society that have begun to multiply rapidly during Perestroika. It is improbable that the newly elected local authorities, let alone the republican parliaments, will be headed by the leaders of environmental movements. More likely, the local governments (Soviets) will be pluralist and the environmentalists will be able to get more than five percent of the votes. In the elections to local Soviets in January to March 1990, 99 percent of the candidates included demands for environmental protection in their election programs. But the local population have, for the time being, different immediate priorities:  provision of housing, food products, medical services, and pensions, and the problems of the old and lonely. Unfortunately, both in Moscow and in local areas, many of the newly-elected deputies understand decentralization of the economy and political power to mean self-sufficiency and even autonomy. They propose that the city, or even individual urban districts, be given complete economic autonomy, with, for example, their own rationing systems and currency.

Finally, as before, the local authorities may possibly be totally subordinated to a major enterprise (particularly in one-plant or one-industry towns, and industrial settlements). A widespread stereotype of mass, and even professional, consciousness is that such enterprises feed the town and that the local inhabitants are its dependents. In this case, local initiative groups face a difficult struggle. As in the West, the success of local groups will depend on how well they are integrated into the wider environmental movement at the state and international level.

## Development and Maintenance of Environmental Initiatives and Movements

### Development of Initiatives

Western sociologists who deal mainly with movements pay less attention to the process of initiative development. But this issue is of high theoretical importance, in particular, because initiatives can and do develop when movements hardly exist. Both in the West and the East initiative groups and voluntary associations are the principal actors in social changes.

The general pattern of development of environmental initiatives is well represented by Nelissen (1991). Nelissen distinguishes the following different types of participation:

Informative participation: administrative agencies (usually after-the-fact) inform the public of specific plans or decisions to be effected;

Contributory participation: governing agencies not only inform citizens of their plans, but also stimulate them to contribute written or verbal reactions to these plans or policy intentions;

Discussion participation: governing agencies not only inform and stimulate citizens' reactions, but organize forums and hearings related to their plans and intentions; and

Decisional participation: citizens also influence the decision making process.

Proceeding from this approach and keeping in mind the importance of the context in which an initiative develops, a more detailed model of an initiative evolution will be outlined. At least ten stages of the process can be distinguished.

First Stage: informative. Residents acquire the right to know about plans and decisions that directly affect them.

Second Stage: responsive. The right to know is augmented with the right to comment publicly on the decisions already taken. But these two flows of information are not interrelated. The procedures for dealing with public opinion and for translating it into the language of design and decision-making are not stipulated. Nevertheless, this step is essential for shaping public opinion within the local community.

Third Stage: discussive. Special procedures and organizational forms are established for dealing with the opinions of residents. Local authorities are pledged to take into account this opinion or even to support and organize special surveys. Mutual understanding and learning processes begin to emerge among residents and local authorities. The "public estimation of urban designs" (Deykov & Varsonovtsev, 1991) is the best example of the study stage.

Fourth Stage: involvement. Residents get involved in the implementation of plans and decisions already adopted. Here residents approach a turning point: being involved in such activity, they begin to realize the ineffectiveness and alienating character of existing top-down, vertical methods of changing their immediate environment.

Fifth Stage: participative. A group participates in decision making. This is the turning point toward the paradigm of environmental self-reliance (self-determination). In particular, it means the beginning of a through transformation of decision-making structures; the shift from top-down to bottom-up decisions.

Sixth Stage: self-reliant initiatives. The major distinguishing features of this stage are emergence of a set of informal groups and their struggle for their own ecological niche in the space that is not controlled by government agencies (the principle of minimal confrontation with the state power). Mobilization resources available include the personal contributions of energy and leisure time of activists.

Seventh Stage: programming (of own activity). At this stage, an initiative group learns how to devise projects and long-term programs. This requires intellectual resources, much more information concerning the state of the local environment, and more deep involvement of activists, including a need for full-time coordinators. Simultaneously, the group looks for backers and sympathizers in the broader social environment. Multilateral informal networks are established.

Eighth Stage: implementation. Programs and implementation of designs developed by an initiative group are begun. The level of mobilization and conflicts grow; achieving consensus with other groups and nonparticipants becomes an issue of constant concern; resources at hand are exhausted; and the group begins to look for both backers and donors to set up its own resource production (money, skills, labor force, raw materials). The initiative group is forced to transform itself into a coordinating center combined with voluntary associations and community enterprises or to affiliate with particular nongovernmental organizations.

Ninth Stage: realization. This stage represents the more or less complete realization of the environmental paradigm. Accordingly, this is the stage at which a new unit of civil society emerges. In more practical terms, what occurs is the formation of an "island" of an alternative way of life and of a new democracy that functions as a carrier and producer of environmental values. The degree of tension between a group (association) enterprise and the local context is at its highest. A group mobilizes all types of resources. The core group seeks contacts with other similar units with a view to mutual support.

Tenth Stage: the beginning of a new cycle of development. Participation in reorganization (of an environment or of decision making) is both the purpose of an initiative and the means for its further evolution. The self-reliant initiative has a variety of ways in which it can proceed: the extinguishing of an initiative; its transformation into a voluntary association (cooperative, community enterprise); transformation into a branch of a public organization; or transformation into a unit of the environmental movement.

## Maintenance and Development of Movements

The first distinctive feature of the East European situation is that the "social movement organization," not to mention the social movement "industry" (McAdam et al., 1988) hardly exists. As they develop, of course, environmental initiatives and movements make increasing use of official resources, i.e., those belonging to the nongovernmental organizations supported by the party-state apparatus. Furthermore, such organizations as the associations of writers, unions of the cinema industry and the artists in the former USSR have made a major contribution at the all-union and republican levels.

The large city milieu inhabited by such giant organizations as government departments is, as a rule, hostile to mass movements. It is important to remember that in East European countries, large bodies of this kind hardly ever compete against each other. On the contrary, they divide the natural and urban environment up into spheres of influence.

Relations between these movements and the state are, at best, extremely strained, but more often confrontational. This is natural because the former try to defend the environment, whereas the latter mainly exploit it. The state has, therefore, achieved minimal penetration of the environmental movement, and the latter acts on the state as an alliance of pressure groups. The state acts as their chief counter-movement because, apart from pseudo-movements set up by the state, environmentalists have no clear opponents.

The media in Russia are highly-centralized and technically underdeveloped, and this seriously holds back the development of environmental movements. Yet, the lack of confidence the movements feel in state-run organizations is more important. They, therefore, prefer to rely on an extensively developed informal network that also penetrates state-controlled bodies. It is this which allows the leaders of such movements, when necessary, to rapidly mobilize large numbers of supporters. The official press and television, particularly at the local level, continue to severely censor the information provided by these movements. Therefore, to begin with, the movements resorted to circulating leaflets, appeals, collections of articles and periodicals in samizdat. In Russia today, certain Green movements already have their own irregular newspapers, and the newspapers and other publications of the independent press also willingly publish their items.

The communication function of environmental initiatives and movements today is to collect objective information. They then pass it on to the public and inform the parliaments of Russia and the other republics about infringements of environmental and other legislation by government departments and local authorities.

As far as the micro-factors of maintenance and development of these movements are concerned, their value system is of primary importance. The similarity between movements in Western and Eastern Europe seems to lie particularly in their aims and values. Unlike other social subjects who strive for the rationalization of the existing economic system or for more just patterns of distribution, the central aim of environmental movements is to preserve the entire world, of which the urban environment forms a part. Their demands are not only directed outwards to political and economic institutions, but primarily at themselves and their immediate surroundings. "If not I, then who else?" "If we don't do something, nobody will" are very typical sentiments for environmentalists.

The emergence of these movements on the basis of interpersonal contacts and the selection of participants for their human qualities points to a common socio-psychological basis for consolidation. Personal satisfaction (inner rewards) is a serious motivating factor. Another integrating value is the capacity to move initiatives and to independently pose and implement specific goals. There is also a definite self-image about the movement, of the defenders of the environment against the intrusions of the state and its institutions.

Environmental values do not simply amount to nature protection and the conservation of resources. Usually the literature refers to an *environmental paradigm* that is the totality of positions held on key problems in society's development: human relation with nature and other people, attitudes toward technology and the economy, as well as the desired principles of public organization and political life (Catton & Dunlap, 1978; Milbrath, 1984; Yanitsky, 1984).

The transition to a new paradigm of social development, presents enormous difficulties, especially because the pressure of extreme living conditions over several generations has drastically lowered the value that Russian culture attaches to nature and human life itself. The waste of resources has become part of everyday life.

At the same time, the Russian people have been prompted by these circumstances to seek a new rallying point. The universal and sharp decline in the health of the environment and the continuous series of ecological accidents (including Chernobyl) have interacted with the lack of values, and a growing feeling of personal oppression and vulnerability. The environmental movements, therefore, might well become the bearers of a changed system of values, i.e., of the new paradigm.

## Summary

The concept of a social movement is no more than a theoretical umbrella which covers a set of events, actions, and processes whose relationships and proportions are highly dependent on the types of actors involved in the given issue, as well as on the types of context in which an individual movement operates. Social movements are generally related to higher, more general goals than are initiatives and collective campaigns (Gusfield, 1981; Pickvance, 1975; Tilly, 1978; Marwell & Oliver, 1984).

We distinguished three types of contexts for the initiatives and movements. Context-I is historical-cultural and includes a given (and stable) system of relations between state, civil society and population, and of basic cultural norms. Context-II is macro-social, and combines stable and innovative elements in the social structure and in cultural norms. Context-III is situational, representing the immediate environment in which initiatives and movements emerge and develop.

We used resource mobilization theory as a primary analytical tool. In general, resource mobilization is not a single act, but a process accompanied by direct and indirect effects. Therefore, the idea of resource mobilization is also a theoretical umbrella which covers a set of different processes. Mobilization is only one stage in the process whereby a movement emerges and then maintains itself.

The production (maintenance) of an environmental movement is asymmetrical. It uses one set of resources (money, labor, knowledge) and produces a quite different set (laws, plans, strategy and tactics, social structures, norms, and values). Therefore, the usual cost-benefit approach to evaluating the movement's results is inappropriate and sometimes senseless.

Public participation can range from passive involvement to self-reliant creative activity. This is also the usual direction of development. Beginning with disbelief and passive observation, people start to doubt and to reflect and then to show an interest in partial involvement. Much later, they turn to independent action. The shift is usually, from reactive and protest actions to creative initiatives, and from partial participation to acceptance of the full burden of responsibility for devising and implementing an initiative from start to finish.

The attitudes of other, external agents to the activities of the population also go through similar stages. At first, there is blanket hostility and even ridicule of local initiatives by the authorities and the planners. Then they begin to consider the novel situation they face and attempt to tame these initiatives. Later, there may be a move towards practical cooperation, although in many instances this stage is not reached. Of course, the rate of this development and, what is much more important, the meaning of each stage, both for activists and for their counterparts, is fully dependent on the context.

Environmental and, in particular, ecological movements are distinct from other social movements in many respects. Environmental movements are a kind of mixture of scientific research and political struggle. These initiatives and movements are, as a rule, long-term or even endless sets of activities. The movements are rather diverse in their social, professional, and age composition; their core is made up of representatives of the human intelligentsia and the academic community who provide the movements with their driving force. Shared human values and long-term cooperation gives these initiatives and movements a community and, even, family character. They are both deeply rooted in the local environment and idealistic in character.

At the beginning of this chapter we posed a set of questions concerning similarities and differences between environmental movements in the West and East. The answers to these questions are entirely determined by the relationships within the state - civil society - population context triangle, because recent grassroots initiatives and environmental movements are part and parcel of modern civil society. With regard to the Russian environmental movements, the following points seem the most important.

In contrast to the West, where civil society is moving into a new phase, conducting experiments in direct democracy, self-management and self-sufficiency, the process of reviving civil society in Russia is still in the initial stage. The residents of Russian cities are still fighting for the right to undertake such experiments, fighting for the very possibility of having a civil society to restore.

Another important point for proper understanding of the processes in question is a vector of ongoing contextual changes. In spite of difficulties and recurrent crises, Western society is evolving toward post-industrial forms. In Russia, after six years of Perestroika, there is still no clear vector of forthcoming changes. Three qualitatively different directions of macro-social context development are being discussed and tested: the revitalization of genuinely Russian (traditional) values; the assimilation of models of Western industrial society; and a search for a third way. Obviously, in each case the role of environmental movements and their local cells will be quite different.

Western civil society, based on private property, a market economy, and civil rights has existed for several centuries. Democratic institutions, formal law, well-developed public opinion, and an extensive network of voluntary associations are habitual and serve as promoters for civil initiatives and the development of new social movements. In Russia, civil society is beginning to revive under the total supremacy of the state, the absence of market economy and many basic civil rights and liberties.

In the West, the state, together with public and private foundations and municipal and local enterprises, maintains grassroots initiatives and assists in shaping the new cells of civil society. In Russia, the contest between the state and these initiatives and cells is growing because they belong to different types of historical-cultural and macro-social contexts.

The most striking contrast, however, lies in the differences in basic social and cultural norms. Whereas the West is consolidated from within a deep consensus based on existing civil rights and liberties, political pluralism, feedback decision-making systems, guarantees of responsiveness to minority opinion, and on individual initiative and responsibility, in Russia the activity of social subjects until now has been integrated by the sanctions of the state, i.e., by the norms of fulfilling orders and depersonalized decisions. The majority of Russian people have no experience in taking initiatives.

Environmental movements, West and East, are the major bearers of environmental values and of the new environmental paradigm, in particular. The difference is that in the West the motto, "Think globally-act locally," has become an element of mass consciousness. In the East, where the ecological situation is much worse, the idea is widespread in the consciousness of urban residents that it is the state which is, above all, responsible for this situation, just as it is also responsible for the destiny of the individual.

The more our comparisons move into the realm of senses and meanings, the more we reveal differences in the social phenomena which, at first glance, seem very much alike. And, obversely, when we turn to the analysis of structural components of the phenomena in question great similarities are observed.

Finally, in the West and in the East, the process of movement development, as described, shows that culture is not an ornament decorating decisions taken by the state or professionals, but a real and vital force that decides the fate of any decision. The reverse is also true: these initiatives and movements are today an important source of the world's cultural development.

# References

Abankina, T. 1986. Empiricheskoje issledovanie vzaimosviazej trudovoj i
  dosugovoj deyatelnosti gorodskogo naselenija [Empirical research on the
  work and leisure activities of the urban population]. In *Problemy razvitija
  sotsialno-territorialnych obsznostei* [Developmental problems of socio-
  demographic groups and local communities] (pp.75-79). Moscow.
Arato, A., & Cohen, J. 1984. Social Movements, Civil Society, and the
  Problem of Sovereignity. *Praxis International, 4, 3*, 266-283.
Arnstein, S. B. 1969. A Ladder of Citizen Participation. *Journal of
  American Institute of Planners*, 35, 216-224.
Bagnasco, C., & Bonnes, M. 1991. Citizens' Participation in the
  Improvement of the Urban Environment in Italy: Development,
  Characteristics and Prospects. In Deelstra, T., & Yanitsky, O., (eds.).
  *Cities of Europe: The Public's Role in Revitalizing the Urban Environment.*
  Moscow: Mezdunarodnye Otnoshenya Publishing House, (pp. 135-149).
Boyden, S., Millar, S., Newcombe, K., & O'Neill, B. 1981. *The Ecology of a
  City and Its People*. Australian National University Press, Canberra.
Bistrup, M., & Svan, M. 1991. The Urban Environment and Public
  Participation in Denmark. In Deelstra, T., & Yanitsky, O., (eds.). *Cities of
  Europe: The Public's Role in Revitalizing the Urban Environment*. Moscow:
  Mezdunarodnye Otnoshenya Publishing House, (pp. 150-162).
Castells, M. 1983. *The City and the Grassroots: A Cross-Cultural Theory of
  Urban Social Movements*. Edward Arnold (Publ.) Ltd; London.
Catton, W., & Dunlap, R. 1978. Environmental Sociology: A New
  Paradigm. *American Sociologist, 13*(2), 41-48.
Catton, W., & Dunlap, R. 1980. A New Ecological Paradigm for Post-
  Exuberant Sociology. *The American Behavioral Scientist*, [Special Issue],
  *24*(1), 15-47.
Cotgrove, S. 1982. *Catastrophe or Cornucopia: The Environment, Politics
  and the Future*. Chichester/New York: Willey and Sons.
Deelstra, T. 1991. Old Buildings and Young People. The Case of Y-Island
  in Amsterdam, the Netherlands. In Deelstra, T., & Yanitsky, O., (eds.).
  *Cities of Europe: The Public's Role in Revitalizing the Urban Environment.*
  Moscow: Mezdunarodnye Otnoshenya Publishing House, (pp. 173-183).
Deelstra, T., & Yanitsky, O., (eds.). 1991. *Cities of Europe: The Public's
  Role in Revitalizing the Urban Environment*. Moscow: Mezdunarodnye
  Otnoshenya Publishing House.
Deutch, K. W. 1985. The systems theory approach as a basis for
  comparative research. *International Social Science Journal, Vol.
  XXXVII*(1), 5-18.
Deykov, A. & Varsonovtsev, D. 1991. Public estimation of project decisions.
  The Bulgarian experience. In Deelstra, T., & Yanitsky, O., (eds.). *Cities*

of Europe: *The Public's Role in Revitalizing the Urban Environment.* Moscow: Mezdunarodnye Otnoshenya Publishing House, (pp. 184-194).

Freeman, J. 1979. Resource Mobilization and Strategy. In M. N. Zald & J. M. McCarthy (eds.), *The Dynamics of Social Movements* (pp. 167-189). Cambridge, MA: Winthrop.

Glazychev, V. 1991. The Research Project "Naberezhnye Chelny" in the Soviet Union. In Deelstra, T., & Yanitsky, O., (eds.). *Cities of Europe: The Public's Role in Revitalizing the Urban Environment.* Moscow: Mezdunarodnye Otnoshenya Publishing House, (pp. 195-211).

Gusfield, T. 1981. Social Movements and Social Change: Perspectives of Linearity and Fluidity. In *Research in Social Movements, Conflict and Change*, Vol. 4, 317-339.

Hackney, R. 1991. Community Enterprise and How to Give Inner Cities New Life. In Deelstra, T., & Yanitsky, O., (eds.). *Cities of Europe: The Public's Role in Revitalizing the Urban Environment.* Moscow: Mezdunarodnye Otnoshenya Publishing House, (pp. 212-235).

Hegedus, Z. 1989. Social Movements and Social Change in Self-Creative Society: New Civil Initiatives in the International Arena. *International Sociology, 4,* 19-36.

Jansson, A. M., & Zucchetto, J. 1978. Energy, Economic and Ecological Relationships for Gothland, Sweden. A Regional System Study. *Ecological Bulletins,* (28), Stockholm, 25-41.

Jenkins, J. C. 1983. Resource Mobilization Theory and the Study of Social Movements. *Annual Review of Sociology, 9,* 527-553.

Klandermans, B. 1984. Mobilization and Participation: Social-Psychological Expansions of Resource Mobilization Theory. *American Sociological Review, 49,* 583-600.

Klandermans, B. 1986. New Social Movements and Resource Mobilization: The European and the American Approach. *Journal of Mass Emergences and Disasters, 4,* 13-37.

Krantz, B. 1991. Shaping Intermediate Level: A Strategy for a New Everyday Life. The Swedish Approach. In Deelstra, T., & Yanitsky, O., (eds.). *Cities of Europe: The Public's Role in Revitalizing the Urban Environment.* Moscow: Mezdunarodnye Otnoshenya Publishing House, (pp. 255-270).

Liddel, H., Mackie, D., & Brown, L. 1991. Participative Projects in Scotland. In Deelstra, T., & Yanitsky, O., (eds.). *Cities of Europe: The Public's Role in Revitalizing the Urban Environment.* Moscow: Mezdunarodnye Otnoshenya Publishing House, (pp. 290-303).

McAdam, D., McCarthy, J., & Zald, M. 1988. Social Movements. In N. J. Smelser (ed.), *Handbook of Sociology* (pp. 695-737), Newbury Park: Sage Publications.

McAdams, R., Smelser, N. J., & Treiman, D. J., (eds.) 1982. *Behavioral and Social Science Research: A National Resource*. Washington, D.C. National Research Council and National Academy Press.

Michelson, W., & Yanitsky, O., (eds.). 1989. *Cities and Ecology*, (Vol. I and II). Nauka Publishing House, Moscow

Milbrath, L. 1984. *Environmentalists: Vanguard for a New Society*. Albany: State University of New York.

Nelissen, N. 1980. *Urban Renewal Participation Experiments. Heralds of a New Local Democracy*? Council of European Municipalities.

Nelissen, N. J. 1982. The Residents' Role in Urban Renewal. *Urban Ecology*, 5(3-4), 305-309.

Nelissen, N. 1991. Methods of Public Participation in Western Europe. Experiments with Public Participation in Urban Renewal in West European Municipalities. In Deelstra, T., & Yanitsky, O., (eds.). *Cities of Europe: The Public's Role in Revitalizing the Urban Environment*. Moscow: Mezdunarodnye Otnoshenya Publishing House, (pp. 53-69).

Oliver, P. 1984. Active and Token Contributors to Local Collective Action. *American Sociological Review*, 49, 601-610.

Parsons, T. 1969. *Politics and Social Structure*. New York: Free Press.

Pahl, R. E., (ed.). 1988. *On Work: Historical, Comparative and Theoretical Approaches*. Oxford: Basil Blackwell, Ltd.

Pickvance, C. G. 1975. On the Study of Urban Social Movements. *Sociological Review*, 23, 24-49.

Pickvance, C. G. 1985. The Rise and Fall of Urban Movements and the Role of Comparative Analysis. *Environment and Planning D: Society and Space*, 3, 31-53.

Pickvance, C. G., 1986a. Concepts, Contexts and Comparison in the Study of Urban Movements: A Reply to M. Castells. *Environment and Planning D: Society and Space*, 4, 221-231.

Pickvance, C. G., 1986b. Voluntary Associations. In R. Burgess, (ed.), *Key Variables in Social Investigation*. Routedge and Kegan Paul.

Saeterdal, A. 1991. Participation as a Learning Process. The Norwegian Context. In Deelstra, T., & Yanitsky, O., (eds.). *Cities of Europe: The Public's Role in Revitalizing the Urban Environment*. Moscow: Mezdunarodnye Otnoshenya Publishing House, (pp. 329-342).

Tilly, Ch. 1978. *From Mobilization to Revolution*. Reading, MA: Addison-Wesley.

Touraine, A. 1976. From Crisis to Critique. *Partisan Review*, 212-223.

Touraine, A. 1981. *The Voice and the Eye: An Analysis of Social Movements*. New York: Cambridge University Press.

Touraine, A. 1983. *Solidarity*. The Analysis of a Social Movement. Poland 1980-1981. Cambridge University Press: Cambridge.

Touraine, A. 1988. *Return of the Actor. Social Theory in Postindustrial Society.* University of Minnesota Press: Minneapolis.

Walsh, E. J. 1981. Resource Mobilization and Citizen Protest in Communities around Three Mile Island. *Social Problems, 26,* 1-21.

Wood, P. 1982. The Environmental Movement. In J. L. Wood & M. Jackson (eds.) *Social Movements* (pp. 201-220). Belmont, CA: Wadsworth.

Yanitsky, O. 1984. Ecological Knowledge in the Theory of Urbanization. *Social Sciences, XV*(2) 133-151. Moscow.

Yanitsky, O. 1986. Chelovetcheskyi factor i sotsialno-vos-proizvodstvennye protsessy [Human factors and the processes of social reproduction]. *Rabochyi klass i sovremennyi mir,* No. 4, 33-44.

Yanitsky, O. 1987. *Ekologicheskaja perspektiva goroda* [Ecological prospects of a city]. Moscow: Mysl Publishing House.

Yanitsky, O. 1988. Obosnovanie gradostroitelnykh resheniy v usloviah glassnosti [Urban designing under conditions of Glasnost]. *Sotsiologicheskye issledovanya,* No. 4, 20-30.

Yanitsky, O., & Glazychev V. 1988. Integration of Social, Economic and Ecological Approaches to Urban Policy and Planning. In *Cities and Ecology* (pp. 58-63). Nauka Publishers: Moscow.

Yanitsky, O. 1991. Environmental Movements: Some Conceptual Issues in East-West Comparisons. In Deelstra, T., & Yanitsky, O., (eds.). *Cities of Europe: The Public's Role in Revitalizing the Urban Environment.* Moscow: Mezdunarodnye Otnoshenya Publishing House, (pp. 81-104).

Yanitsky, O. 1991. Fight for Lefortovo, Moscow: Resolving the Conflict between Urban Planners and Residents. In Deelstra, T., & Yanitsky, O., (eds.). *Cities of Europe: The Public's Role in Revitalizing the Urban Environment.* Moscow: Mezdunarodnye Otnoshenya Publishing House, (pp. 356-371).

Yanitsky, O. 1993. *Russian Environmentalism: Leading Figures, Opinions, Facts.* Moscow: Mezdunarodnye Otnosheniya Publishing House.

Zald, M. N. & McCarthy, J. D. (eds.). 1979. *The Dynamics of Social Movements.* Cambridge, MA: Winthrop.

Zald, M. N. & McCarthy, J. D. 1987. *Social Movements in an Organizational Society.* New Brunswick, NJ: Transaction Books.

Chapter 7

# THE STRUCTURAL MECHANISMS OF THE ORGANIZATION OF ECOLOGICAL - SOCIAL MOVEMENTS IN HUNGARY

Viktoria Szirmai

Pictures of the tennis players playing without a ball, the closing scenes of Antonioni's famous film, *Blow Up*, occur to me when I want to analyze the environmental movements of the 1980s. The people in the tennis court of the film do not need a ball, as only the imitation of the game is important and the ball would only disturb it and make it serious. The natural environment and the endangered inhabitants were not really present in the protective "game" of state socialism either. In the 1970s popular movements of environmental origin arose only occasionally in various locations of Hungary. According to the sociological research of the period, even these movements did not occur for ecological and health considerations, but for other reasons (Szirmai & Lehoczki, 1988). Those who were always on the scene, such as the environmental protection authorities, the ministry, the councils, and even the large enterprises, did something for ecological issues only apparently, occasionally, and indirectly.

Prior to the political system's change, the population and the movements representing them, made to sit at the bench by the court, appeared not only to become the most important players, but also to demand, at last, the much missed ball. The facts show, however, that only the rules of the game had changed: there is still no ball and only the imitation of the game is important. The movements seem at the baseline of the court again. This became apparent from the fact, among others, that the number of spontaneous ecological movements, which had increased significantly in the 1980s had dropped considerably following the transformation of the political system.

The number and geographic distribution of ecological movements has not reflected the critical condition of the environment in Hungary. What is the reason for the historically changing significance and fundamentally limited role of the various ecological movements; what was the role and function of the ecological movements in the 1980s and what is it now; and what are the most important reasons for these? In this paper I try to find an answer to these questions.

*A. Vari and P. Tamas (eds.), Environment and Democratic Transition, 146–156.*

## Environmental Protection in the State Socialist Structure

The ecological risks have not always been obvious even in advanced industrial societies. The structure of economic, political, and social interests and the requirements of the rapidly unfolding economic development following World War II did not integrate considerations of environmental protection (World Committee on the Environment and Development, 1988). This is one reason why economic progress, modern urbanization, and the growth of cities have been accompanied by significant environmental damage. Since the mid-1950s living standards have risen spectacularly and the quality of life has improved in Western Europe and in the United States. Production growth and new technology have not only brought about significant development, but also resulted in an intensive utilization of raw materials and energy, resulting in the destruction of the environment.

During the temporary economic recession of the 1970s it became obvious that environmental problems hampered and retarded economic development in the advanced industrial countries. Social dissatisfaction was growing as a consequence of ecological problems. Social movements and alternative ecological movements were organized in the United States, Japan, the Federal Republic of Germany, France, Britain, and the Scandinavian countries in the 1970s, and in the South European countries in the 1980s. The social movements, the growing demand of the middle classes for environment-friendly products encouraged the industries to develop technologies and industrial processes suited to reducing adverse ecological consequences. The political force of the ecological movements and the green parties organized in some countries has led to the evolution of environmental protection by the state.

In the East European societies, including Hungary, all those conditions were missing which had led to the evolution of environmental protection by the state in the advanced Western societies. The highly concentrated economic structure and the forced industrialization launched in 1950 and oriented towards mining and heavy industries overburdened the natural environment. The centralized political structure ignored ecological considerations because of economic and ideological interests. In the 1950s and 1960s, the practice of planning and investment built up a multitude of new towns and industries without even thinking about the impact upon the natural environment. Even the official ideology expressed the disadvantages of ecological considerations because the symbol of the hoped-for socialist way of life was the large factory and its tall chimney, usually located between housing estates, next to the town. The political system did not give preference to the social forces interested in the protection of the ecology.

The alternative movements were banned. Because of the dependency of individuals on central authority, no horizontally organized groups or movements could evolve, even spontaneously, in the hierarchized model of social planning organized from top to bottom. Organization and movements were not possible because people were totally uninformed about the ecological issues which had been centralized and made secret. Not only the power structure and ideological interests of the political system were hidden behind the model, but socialist industry was unable to produce the resources necessary, for environmental protection. The driving forces of the market and the society were missing.

For a long time the state socialist political system tried to separate the country from those processes that had been unfolding with growing intensity in the field of international environmental protection in the advanced market economies from the late 1960s onwards. The party documents of the 1970s reflected the idea that in socialism no environmental protection of autonomous interests was needed because each actor in the socialist economy protected the environment automatically, under inherent pressures.

## The Civil Society

In the advanced market economies the most important forces promoting environmental protection are the civil society, that is, a bourgeois middle-class that demands environment-friendly technologies and products, and those social movements and ecology-centered political forces which are either less or not at all integrated into capitalist society. In East European socialist state structures there is no civil society and no bourgeois middle-class because of the backwardness of bourgeois development, the constant dependency upon big powers, and the lack of demand by the population influencing production. In Hungary, it was particularly the period following the communist takeover in the early 1950s which destroyed the historical results of the erstwhile organic bourgeoisie. As a result of the political and economic modernization of the 1960s and 1970s, however, a quasi-market evolved under state control, the so-called second (but not black) economy, which produced the possibility of a relative economic autonomy and the setting up of individual life strategies. As a result a process of petit bourgeois growth occurred: people bought flats and cars and started to build a second home. A "second" and not civil society evolved partly beside, and partly linked to the "first" social sphere controlled and formalized by the state. The second society has preserved not only the bourgeois traditions rooted in the historical past, but parallel to the weakening of state socialist authority and centralization, it has begun to build the elements of a modern civil society.

Broad ecological interest did not develop as a consequence of all this. The process of petit bourgeois growth was sufficient only to satisfy the most important primary needs, and that, too, with much labor. But bourgeois growth did not favor the expression of post-materialist values, nor the demand for environment-friendly products. Because information related to ecological damage had been centralized, people continued to be ignorant about the ecological problems. The system of central dependency survived, although the decentralization of the state socialist structure devolved it to the county, or local authorities. The local government, theoretically interested in environmental protection, was controlled by the local political authorities, whereas the latter were exposed to the central will, not giving preference to environmental protection.

## The Role of the Intelligentsia

The international news about global environmental problems was first brought to Hungary mostly by scientists, experts on ecological issues, and intellectuals participating in various international conferences. During the course of the 1970s an increasing number of Hungarian experts dealing with ecological problems -- biologists, chemists, urban and regional planners -- discovered environmental damage in their own fields of specialization. The effect of the influence of the specialists has been highly significant, as it was among them that ecological thinking and the practical implementation of ecological considerations were first organized without central initiatives. In their immediate environment, medical doctors, educationists, and kindergarten nurses who were informed by foreign literature, began to see and experience with growing frequency, the adverse effects upon health in their localities. These effects were caused by industrial and agricultural environmental damage. Many of these professionals called attention to the environmental problems of nature. Some groups of Hungarian society, such as residents of some contaminated settlements, called attention to such problems mostly with the help of the mass media. Whenever they could they mobilized the authorities and the concerned inhabitants as well. The role of public administrators, and other public policy makers, cannot be underestimated either. They did not and could not publicize the results of the studies of the environment kept secret by the state's socialist centralization of information and by the lack of social demand, yet they could informally leak the facts and let them reach physicians, the opposition, and even the inhabitants concerned. This is how a number of the local ecological conflicts of the 1970s and 1980s were prompted. This is how in 1965 the first public scandal broke out about the decay of fish in Lake Balaton, and how it became public that the waters of River Sajo were dying. It was at that time that bathing was prohibited in the Danube because of pollution.

The intellectuals interested in the development of environmental protection by the state, made it clear to the political authority by informal means that the institutionalization of environmental protection was necessary. From the 1960s and 1970s onwards. The central political authority was forced to yield and to strike a bargain with the professional groups involved for two reasons. First, the power centre, interested in Western relations and credit, had to pay attention to its international image. Although the policy of exclusivity was still continued as late as the mid-1970s, the central authority agreed to establishing a state environmental protection agency of narrow competency and modest opportunity for representing environmental interests. The second reason was political fear of the massive spread of ecological movements. This might be another reason why a politically controlled, yet resource-consuming system of environmental protection was decided upon.

The environmental protection branch, utilized the opportunities of the armory at hand. It drew strength and basis from the second society in order to acquire a greater autonomy, authority, or legitimacy. Part of the scandals related to the dissatisfaction of the population due to the environmental damage and pollution did and could break out because the environmental protection branch wanted to prove its strength. By supporting local protest movements, it could achieve a greater power of interest assertion and larger resources for development.

In the 1970s there were only sporadic local ecological conflicts. In January 1974 the 3000 cubic meters of fuel oil released by the Danube Iron Works into the river caused panic in Dunaujvaros. In 1978, the population of Nagyteteny (Budapest) was outraged because of the series of illnesses caused by the lead contamination by the Metallokemia firm.

In the 1980s an increasing number of ecological conflicts broke out. In 1980 a series of reports by the inhabitants was sent to the Public Health Authority of Vac claiming that the wells and pipe water were polluted by inappropriately stored hazardous wastes of Chinoin. In January, 1984 the workers of the glass factory of Ajka protested against the dust pollution emitted by the chimney of the Ajka Thermal Station. In November, 1984, a strong local resistance developed against a planned hazardous waste incinerator in Dorog.

From the 1980s onwards, parallel to the growing of the command economy, it became increasingly difficult to obtain larger sums of money from the state. Environmental protection, however, remained a priority for the central government. On this basis curious alliances of interest were built. Central support for environmental protection was useful for the natural environment. It was good for the local authorities as well because the crisis of legitimacy related to the dissatisfaction of the population with environmental damage was reduced. And it was good for some large enterprises incurring losses, as the money received for environmental protection -- even if it were not used for something else -- although there were innumerable examples of it -- temporarily offered a breath of air to the leaders. The controversial alliance of industry and government was shared by the workers of the polluting factories too, who, as the inhabitants of the town or part of it, suffered from the environmental damage, but who, as workers, were interested in maintaining the factory. From their perspective the damage caused had not been very dangerous and the investment into environmental protection had already solved the problem. Finally, the given system was good for the environmental authorities, too, as they could prove that their activities were directed towards environmental protection, whereas they did not violate the basic political rule, namely, that popular dissatisfaction should not be whipped up, but rather controlled, and the interests of the large enterprises should not be touched. But if at any moment the agreement was overthrown -- that is, if money allocated for environmental protection had been reduced -- the central authority could be told that both the institutions of environmental protection and the loyalty of the local authorities would cease.

## Alternative Ecological Movements at the Threshold of Systemic Change

In the 1980s the processes causing the complex crisis of the state socialist social structure were accelerated. The existing relationship between the first and second societies was shaken. The spontaneous environmental movements gained strength with the increased number of newly-founded associations which were created primarily in response to local environmental problems (Solyom & Szabo, 1988). The most important of them was the Danube Movement, which was organized in response to the anticipated adverse ecological consequences of the Bos/Gabcsikovo/Nagymaros dam under construction in 1984. This movement was not a local one anymore. The case got out of the sphere of the Budapest intelligentsia. A letter was sent to Parliament and to the government with more than 10,000 signatures demanding the immediate suspension of construction and the elaboration of such a plan as would pay attention to the ecological consequences, too. This movement did not resemble the earlier protest actions. It was not linked in any way to the agreement on interests which had existed between the first and the second societies. Its unfolding was not inspired by the institutional system of state environmental protection. The movement attacked the social and political structure which was the framework of environmental protection provided for by the state. It attached the centralized redistributive system where an increasingly intense struggle was going on for existing resources; and environmental protection could have access to the resources, only if it supported such major economic organizations which caused environmental damage but which would also guarantee the central development funds necessary to the reduction of that very damage (Fleischer, 1988).

In October, 1985 the Danube Movement was awarded the alternative Nobel Prize, the Right Livelihood Award. The alternative Nobel Prize was not only for the recognition of the movement but also of the role it performed in social transition. After the appearance of the Danube Movement, local and regional ecological movements had to be taken more seriously as negotiating partners.

Opposition to nuclear power, generally characteristic of the ecological movements, was missing in Hungary. In 1976, when the Paks nuclear power station was built, no protest demonstration was organized against it. In the late 1980s however, the population of Ofalu of Baranya County not only revolted against the plan to build a permanent low-level radioactive waste disposal facility, but effectively halted it. (Juhasz et.al., this volume). Thus the strength of the ecological movement was again demonstrated.

## Ecological Movements after the Political Transition

New political parties founded just before or after the dissolution of one-party rule included ecological programmes in their election promises, but the ecological issues were pushed into the background with the establishment of the new political and power structure. The processes of party-building have proved to be stronger than the organization of movements, whereas the new parties have sucked in the representatives of the green cause (Gallo, 1991). There had been many people among the environmentalists who participated in the ecological movement because of direct political considerations under the state socialist political regime because the erstwhile authority was somewhat more tolerant towards protests over environmental issues than in the case of unambiguously political actions. Politically motivated members of the movement have become members of different parties, often remaining within the green groups of the respective parties after the change.

The ecological issues are included as problems in the programme of the government, but there has been no concrete concept of ecological activity as yet. In keeping with governmental interests and those of privatization, state environmental protection has been retreating from some ecologically important fields, such as the cleanup of several contaminated facilities. The government mostly deals with environmental issues and represents them only if their different international, but mainly political, and partly economic, interests so require.

The Parliamentary committee dealing with ecological issues cannot change much of the existing situation. A series of in-depth interviews done among those involved suggest that the opposition MPs have little information on ecological problems because efforts towards centralization have been recurrent, and it is still difficult to get access to environmental data (Gallo, 1991).

The possibilities of the assertion of the ecological interests of local governments are rather limited. The new acts and rules regulating property rights have not yet been prepared, so it is impossible to develop local protection, particularly when state environmental protection is being withdrawn and is not accompanied by the development of conditions by the state which would safeguard local environmental protection. New constraints emerge because of the great shortage of resources and new dependency of locally-polluting factories and large enterprises upon the regularly currently continuing privatization in which ecological interests are swept aside. The strong party dependence of the governments also works against ecological considerations. Quite commonly, even on the local level, ecological questions are only discussed if they are so required by party interests.

In November, 1990 there were 54 political and social organizations of ecological nature in Hungary. They number much more than ever before. Yet the situation of the ecological movements is difficult. They have been squeezed out of the center of political life, and primarily they aim at influencing, with varying success for the time being, local government decisions related to environmental issues.

## Summary:  The Role of the Ecological Movements and Ecological Interest Assertion

The significance of the local social initiatives and ecological movements in changing the political system in Hungary is beyond doubt, particularly during the second part of the 1980s and early 1990s. The emergence of the spontaneous movements of the 1970s, however, and even of the late 1980s resulted in the temporary strengthening of the state socialist structure, just as the calming down and silencing of the ecological movements during the period of the bourgeois democratic transition also caused the consolidation and stability of the political system.

In the late-comer and backward Hungarian modernization, the state (state party), was hardly interested in environmental protection. The changing but always narrow possibilities for interest assertion of the public resulted in different forms of environmental protection, including the movements serving social, structural, and mostly political interests rather than the considerations for the environment.

In the state socialist structure there were rational causes for ecological issues being utilized for diverse interests. First, those with local power, depending upon state and monolithic party institutions, could have access to sources of development and could legitimize their power. Secondly, it was not simply due to the weakness and backwardness of civil society, but rather to its defensive reflexes, and to the contradictory structure of interests related to environmental damage that the inhabitants of the polluted regions were simultaneously -- although to different extents - interested in liquidating pollution and in preserving the polluting enterprise, the workplace. The processes of bargaining between the first and second societies and the temporary stabilization of the political and economic system derived from this contradictory structure of interests.

Neither the ecological interests of the social structure, nor the historically conditioned backwardness of Hungarian efforts to modernize changed after the political change. The greatest problem is caused by the lack of resources available for environmental protection and by the financial difficulties of the government. For the time being, the Hungarian economy has been unable to produce the resources necessary for environmental protection, to develop an economy and industry interested in the protection of the environment and to develop adequate technologies even under the conditions of transformation. In this sense, the ecological behavior of the government, as well as of the opposition parties is a rational one. The political parties act rationally when they advocate ecological objectives in their election promises, because they know, or guess, that a promise to correct the damage will get votes.[1] As the acts and economic means necessary for real action are missing, they do act rationally even when they push the ecological issues into the background after having acquired power. The ecological interests of society and of governments, of the voters and potential voters, continue to be divided. A solvent middle class demanding environment-friendly products is still missing in Hungarian society. Interests related to the daily needs of life, to the place of work and livelihood are apparently more important than the environmental problems. Today it is more important for a miner, or to the inhabitants of a town organized around an aluminum foundry to know what would happen to them and to their families if the government closed down the mine or the aluminum foundry, than to know about the risk of disease. In such situations the local government is bound to fail if it represents the ecological and not the general chances of the livelihood of the town.

Lack of interest in ecology is also explained by the fact that some people have thought or wanted to think that the already legitimate Parliament, central government and local governments would look after the ecological issues. The facts, however, show another trend as well: the germs of a new type of ecological movement. The realization, if not a direct ecological sensitivity, is growing among those who have been squeezed out of the political system, or have been in a more peripheral power or social position, that the ecological problems continue to offer an opportunity to satisfy certain local economic, political, and power interests. Local communities often turn involuntarily, or under the pressure of an alert to the means of asserting ecological interests, particularly if their daily life, work, material and income conditions are endangered. Concrete examples of this are the newly unfolding local opposition against the siting of a low-level radioactive waste repository in Kovagoszollos, which is a local effort in the face of the interests of national authorities and of a large enterprise.

These movements already represent a new phase of the assertion of ecological interests despite the formal similarities with the earlier movements.  These movements are not aimed at challenging the state socialist setup and legitimizing bourgeois transformation.  These forces do not want to accept the present situation where the ecological decisions (as well as all others) are made without a broad social consensus, but in bargaining processes among the ruling parties and the government.  This also proves that the ecological movements may have a great variety of roles, and are linked to different interests, at times only touching on the ecological factors.

The role of the ecological movements, citizens' initiatives, and protests is significant in Hungary, and their history is similar to the history of environmental protection in Western societies.  Historically speaking, they promoted the evolution of environmental protection by the state, and, subsequently, its progress towards modern Western environmental protection measures.

The Hungarian movements of the 1980s laid the foundation of the development of environmental protection in a bourgeois democracy and did not demand democracy beyond the limitations of a bourgeois democratic setup (Szabo, 1989).  This is where they differed from the objectives of the Western alternative movements, which wished to go beyond the bourgeois setup.  The difference between the Hungarian and the East European ecological movements, in general, and the international alternative movements is due to the different phases of history and challenges for modernization.  The Eastern and the Western movements face different social, economic, and political problems because of the different histories of modernization.  The Western alternatives seek a way out from the crisis of an advanced market economy, whereas the Eastern see the possibility of emergence from a state socialist crisis by emulating an advanced market model.

## Notes

1.  As shown by a research study, the expectations of the politicians are not wrong. For instance, the inhabitants who had been interested in closing down a polluting factory in a peripheral district of Budapest, who lived near the factory but did not work there, voted for the candidate -- of the governing party -- promising a solution, even if otherwise they had agreed with the program of other parties (Berki et al., 1990).

# References

Berki, S., Szilagyi, K., Bartfai, E., Hasko, K., & Gallos, J. 1990. *Vezetok es kornyezetvedelem (Leaders and environmental protection)*. Case study, commissioned by the Ministry of Environmental Protection, done by the Institute of Sociology, HAS, Budapest. Manuscript, p. 229.

Gallo, B. 1991. *Halvanyzoldek (Light Greens). Pillanatkep az okologiai jellegu es tarsadalmi szervezodesek helyzeterol egy evvel az 1990-es marciusi valasztasok utan. (Snapshot of the situation of ecological and social organizations one year after the March 1990 elections.)* Budapest. Manuscript, p. 17.

Fleischer, T. 1988. Nem kell mindennek kifizetodonek lenni, de... (Everything need not be profitable, but...) *Magyar Nemzet*, September 17.

Juhasz, J., Vari, A., & Tolgyesi, J. (This volume). *Environmental conflict and political change: The case of low-level radioactive waste management in Hungary.*

Persanyi, M. 1986. *A kornyezetvedo allampolgari tevekenyseg hazankban... (The development of the activities of citizens and movements protecting the environment abroad and in our country)*. Ph.D. dissertation, Budapest.

Solyom, L. & Szabo, M. 1988. *A zold hullam (The green wave)*. ELTE Faculty of Law and Political Science, Budapest.

Szabo, M. 1989. *Politikai okologia (Political ecology)*. Bolcsesz Index, Central Konyvek 2., Budapest

Szirmai, V. & Lehoczki, Zs. 1988. *Kornyezetallapot es erdekviszonyok Ajkan (The condition of the environment and interest relations at Ajka)*. Case study, commissined by the Ministry of Environmental Protection. Budapest, Manuscript, p. 148.

Szirmai, V. 1989. *A kornyezetvedelem erdekviszonyaui es a kornyezetvedelmi politika (Interest relations and the policy of environmental protection)*. Summary paper. Budapest. Commissioned by the Ministry of Environmental Protection. Vols. I-II. p. 150.

World Committee on the Environment and Development 1988. *Kozos Jovonk (Our common future)*. Mezogazdasagi Kiado, Budapest.

Chapter 8

# ENVIRONMENTAL PROTEST AS A VEHICLE FOR TRANSITION: THE CASE OF EKOGLASNOST IN BULGARIA

Bernd Baumgartl

In this research paper,[1] the conditions for the foundation, structures and consequences of the ecological movement in Bulgaria are analyzed. Most studies on the environmental situation in Eastern Europe have concentrated on the USSR or on Central Eastern Europe (Eastern Germany, Czechoslovakia, Hungary). Bulgaria has often been left out and somehow remained unnoticed in this respect. On the other hand, the particular importance the environmental issue gained in the Bulgarian transition justifies a scrutiny of the Bulgarian case. At least for a certain period of time, the ecological movement was the only possibility to express political discontent and opposition. Ekoglasnost was one of the first national movements under the Communist government in Bulgaria. It started as a protest to improve the environmental situation in the town of Ruse, in northern Bulgaria. This town had been (and still is) seriously polluted by a Romanian chemical factory on the other side of the Danube in Giurgiu. All attempts to change the situation failed. The Bulgarian authorities did not take care of the problem. After the foundation of a Committee to Save Ruse the issue gained national interest. The successor of this solidarity group named itself Ekoglasnost and groups expressing their support emerged all over the country. Further stimulated through political repression, the ecological movement shifted to broader political issues (such as the human rights discussion). Within the transition process in Eastern Europe, Ekoglasnost took a decisive role in Bulgaria.

The emphasis of this study is to analyze the changes within the ecological movement, starting with the Ruse developments in autumn 1987, and ending in spring 1990, on the eve of the first free elections in about 45 years. The question of why, precisely, the environmental issue served as the unifying theme cannot be answered without a view of the political opportunity structure for opposition to the Communist government and the characteristics of the environmental theme itself.

The influence of the political opportunity structure on the forms of action which Western social movements adopt has been analyzed by Kriesi (1990) - among others. He demonstrated that the number of institutional channels to express political disagreement defines the grade and intensity of social action. At a certain level of political repression, protest seems to cease. Certainly the non-existence of ecological protest until a few years ago is also partly due

A. Vari and P. Tamas (eds.), Environment and Democratic Transition, 157–175.

to an extremely hostile political opportunity structure. When the growing success of Gorbachev's perestroika was spreading to other countries, Bulgaria remained more conservative than the USSR, which had always been held up as an example to follow. As control of the party over society became weaker, it was merely a question of time, and finding the theme around which first opposition could be formed.

In Communist Bulgaria, there were no institutional channels for political opposition which could accommodate opposition outside the system. Partly as a strategy of the founders, and partly because of its characteristics, the environmental issue became the vehicle for the first steps of the Bulgarian transition.

## Preliminaries of the Green Movement in Bulgaria

### Communism and Ecology

At the United Nations Conference on the Environment in Stockholm in 1972, the Socialist countries called environmental damage a capitalist problem, a statement that seemed to have a great impact on Western judgments of the environmental awareness in these countries (Weizsacker, 1989; Spetter, 1988). The socialist view that environmental problems were capitalist and, therefore, Western problem has been noted by many authors (Kaloyanoff, 1971; Kelley, 1976; Spetter, 1988; Redclift, 1989).

Environmental management under Communist governments can be described rather briefly: "It is the history of a failure" (Jancar, 1987).

Until the end of the 60s, the Soviet Union and the rest of the socialist bloc dismissed environmental problems as the evil results of capitalism. Citizens were assured that the socialist system, particularly the public ownership of production, prevented such abuses from occurring in the socialist countries. The eruption of national and then international scientific concern over ... [specific issues] brought recognition that the USSR, like capitalist nations, did have pollution problems, but that these problems were being taken well in hand by the party and government (Jancar, 1987).

The dogmatic assumption that environmental damage could never be caused by a state, that has no profit interests, prevented an effective tackling of the issue while the damage was still less serious. Nevertheless, early warnings about the consequences of industrialization were pronounced in Socialist countries. Czechoslovakia and Poland even started to look into pollution in the late 1960s, and in 1971 they noted that "widespread concern is becoming apparent from the Baltic to the Mediterranean" (Kaloyanoff, 1971).

Still, in the 1970s, some idealistic theoretical Western scientists hoped that a nationalized economy would be easier to regulate than Western private enterprises (Kelley, 1976). Subsequent information proved that the environmental situation in Eastern Europe was even worse than in the West. Socialist productivity was far lower than capitalist levels, but consumed one-third more energy (Oschlies, 1987). Especially during the last 15 years, the environmental situation has deteriorated rapidly (Davy, 1990).

Lots of explanations have been given for the poor state of the environmental situation in Eastern Europe including the lack of attempts to assess externalities; shortcomings in cost-benefit analyses; and emphasis on gross national products and economic growth, rather than the concept of the net national well-being (Goldman, 1972). Efficient control was prevented by the planned economy itself. Often a single authority was responsible for both the pollution and the prevention. Even if polluters were detected, it was a lot easier and cheaper not to stop the source of environmental damage, but to pay the disproportionately low fines. According to a report of the official newspaper of the Bulgarian Communist Party,[2] in 98 percent of the cases that were brought to the courts, polluters were not fined. Their judgments were based on the fact that managers had circumvented environmental laws in order to fulfill their plan targets; this was considered a more important goal (Spetter, 1988).

According to Ashley (1988) Bulgarian environmental law was not enforced because of lack of coherence. Contradictory laws on plan fulfillment, the poor quality of the purifying installations, and managers who acted with the best motives to meet plan targets.

Lack of information may partially explain the absence of ecological concern. A considerable body of the ecological data was held back, as if it were a state secret. Information at all levels was unavailable: One could not even ascertain the level of pollution or assess whether suspected environmental damage was dangerous. Nobody would provide information or identify the person or authority in charge of the issue. There was no possibility of claiming an ecological right. Even experts with access to confidential information were unable to rely on it, because official statements often turned out to be false.

A satisfactory analysis of the failures in environmental spheres cannot be addressed in this paper, and this is not the aim. It is sufficient to state that the lack of responsibility on all levels turned the supposed advantages of the system into disadvantages. The state was the only institution in a position to cope with ecological issues and it fails to do so, and there was no power left to control the state.

As a contrast to the European Community, which, at least, from 1987 on, started to act also on an interstate basis,[3] the Council for Mutual Economic Aid did not regulate environmental improvement in a transnational framework (Kusin, 1987). Concerns with economic growth continued to tower above all other recognitions. Eastern Europe remained clearly behind the West in protecting the quality of life.

The concept that energy, water, and soil should be available free for citizens, (or, at least, at the lowest possible prices) had tremendous ecological consequences. Energy waste was extremely high and not fined at all. The symbolic images of working machines and producing factories ensuring the well-being of the population, and of smoking chimneys, reflecting a quickly modernizing society, indicated an overall attitude closely linked to the ruling Marxist ideology and political theory which emphasized economics, industrialization, and technical progress.

## Environmental Protection in Socialist Bulgaria

The first public discussion on environmental issues took place as early as 1960. Press articles concerning waste-water were followed by the first legislation on the subject, a decree on the protection of nature. A second law in 1963 on "Air, Water and Earth" demanded purifying systems, but was never enforced (R.T., 1972).

In 1970, a Western newspaper reported that 80 percent of Bulgaria's water sources were contaminated, and that 50 percent were so heavily polluted that the water could not be used for irrigation.[4] Ninety percent of the enterprises using water reportedly had no water-purifying equipment at all. In addition to water (and subsequently soil) pollution, air pollution and noise were already identified as serious environmental problems.

In 1971 a Ministry for Woods and Environmental Protection, one of the earliest such in the world, was created. The establishment at about this time of additional institutions such as a Committee for the Protection and Reproduction of the Natural Environment (a council inside the State Council), the Scientific Coordination Center for the Protection and Reconstruction of the Environment (an institute inside the Academy of Sciences), and a National Committee for Environmental Protection (inside the Patriotic Front), seemed to reflect a high degree of awareness of ecological problems (Oschlies, 1985). Nevertheless, by this time 40 percent of all Bulgarians lived in air-polluted areas.

Successes were enthusiastically reported as early as 1973 with diminishing air pollution in the area of Pernik, an industrial town southwest of Sofia.[5] In 1977 a decree of the State Council stated that the main aim of the ecological policies was:

The preservation and improvement of the quality of the natural environment as a basic source of resources and a vital milieu for the present and the future generations, and setting forth fuller harmonization of the interrelationship between society and nature.[6]

The problem obviously was not a complete lack of environmental directives, but, rather, the enormous number of often contradictory decisions and their implementation. As an example, in order to stop the pollution of the L. I. Breznev Steel Works (formerly and recently again Kremikovtsi) near Sofia, 150 decrees and orders were issued by the government. Nevertheless, this factory continued to emit hundreds of tons of dust and sulfur dioxide daily (Ashley, 1988).

Many other examples of extremely dangerous pollution can be cited. Most shocking for the broader public was the result of an experiment carried out at a bridge (Orlov Most) in the center of Sofia in 1990; a hamster died after four hours because of polluted air (Ekimova, 1990). This occurred in a country where 67 percent of the population lives in towns. Not only the capital, but also several provincial towns, had to fight with extreme transgressions of the permitted standards.

Water and air pollution was accompanied by increasing levels of soil erosion, which diminish productivity (Oschlies, 1985). Erosion by wind (which affects 30 percent of the cultivated area) and water (endangering 80 percent of the agriculturally used ground-space) is due to an intensive exploitation of the soil with single-crop farming and oversized machines. Because of its extremely poor water resources, the pollution of the subsoil water poses a more urgent danger in Bulgaria than elsewhere.

## Evolution of Ecological Discomfort

The preoccupation with the concept of nature and the relation between man and nature in the Bulgarian literature has gone through a remarkable change of emphasis. During the 1960s, nature was still seen as an obstacle to the achievement of technical progress. Poetry of this era, thus, called for an acceptance of technics. Later, poets began to criticize particular projects in which the environment was damaged locally. The pollution of Ruse was first made public by poets in literary journals. Poets thus played the role of an avant-garde, as sensitive observers of society, and challenged the legitimacy of an industrial-instrumental use of nature, a pattern which has also been observed in Western societies by Lauber (1988). As industrial progress comes to be seen as a consequence of the predominant ideology and the scientific-technical revolution, the leading role of scientists is questioned. A naive "belief in science" is to blame for neglecting awareness of nature. A further stage of criticism is to doubt the socialist ideology of progress and welfare which is to be achieved through industrialization, but which is soon enlarged by a transnational and transideological critique and fright. At the end pessimism provokes cynicism about human hubris. Blaga Dimitrova asked in a poem of 1988: "Dokade ste stignem?"[7]

Even if such ideas opposed the rationality of Communist values, it was difficult for the Communist Party to complain about the goals environmentalists expressed. The ecological issue was not in direct confrontation with the party line. The government presumably was also committed to improving the environment (Fisher, 1990). Moreover, on a theoretical level, Communists claimed to share these goals as a pre-condition to ensure a better life for the population.

The visible effects of environmental damage hindered the party in fighting environmentalist actions in the same fierce way it had fought other protest movements. The consequences of ecological criticisms endangered the party's monopoly, however, so it could not tolerate complete freedom of action for the green movement. The growing support for green movements is explained by Vargha (1990) by the fact that public opinion in Eastern Europe linked the crisis of the environment with the political system. The impossibility to suppress the independent groups made the government try to establish counter committees and organizations.

Also significant, the original, particular cause for the beginning of protests was reportedly a Romanian factory. Therefore, a foreign polluter was to blame, and Bulgarian authorities were not accused directly. To a certain degree, it was possible for party members and organizations to declare their solidarity with the ecological protest. Although the pollution of Ruse was not caused exclusively by this foreign firm, the existence of an external enemy played a key role in the party's reaction. The impossibility of interfering directly in Rumania's internal affairs may have influenced the party oligarchy to underestimate the potential power of this movement.

## The Evolution of the Green Movement

The environmental organizations founded by, or under the auspices of, the Communist Party were not truly concerned about broad ecological problems. Most of them served more as an alibi than as criticisms of environmental damage. Some of them were related to the National Parks or the organization of alpinism. The independent opposition environmental groups which preceded Ekoglasnost were never able to overcome the initial problems resulting from governmental repression.

### The Ruse Committee

Ruse is the fourth largest town in Bulgaria, situated at the Northern border of the country. Air pollution in this town had reached extremely high levels, caused mainly by a Romanian chlorine factory on the other bank of the

Danube in Giurgiu. This plant, producing caustic soda was installed in 1984 and was equipped with Russian technology. There was little information about the proceedings and output of this production, and none in the Romanian press. From 1986/87 on, probably because of inefficient maintenance, the already existing pollution increased heavily. The Bulgarian town of Ruse was badly affected because of the predominating wind from the north. Still, in the era of Zivkov and Ceaucescu, criticisms, were answered with the comment that Russian technology`s failures should be the responsibility of the Russians themselves. Russian officials maintained that this type of chemical factory would work without damage to the environment in the Soviet Union. In 1986 there were even plans to expand the production, which was destined for export and, therefore, an important source of hard currency.

The question of Ruse had been raised first in 1984 by a regional newspaper, the *Dunavska Pravda*.[8] The tendency of the articles was clearly to understate the extent of the problem. In 1987 alone, dozens of gas emissions occurred, up to nine times higher than the permissible limits. Two particularly large gas emissions on September 23 and 25, 1987, caused the first uprising and demonstrations in Ruse on September 28. A letter was published, signed by "thousands of mothers of Ruse":

We want this [chemical] plant to be closed down...this is the demand of all mothers who want to save the lives of their children...Are you going to create another Chernobyl or Bhopal?...We want to live and work normally, without fear...There are enough miscarriages, enough stillborn children. Stop it!...Our patience is exhausted (G.S., 1988/b).

The Ruse autumn (Esenta na Ruse) continued with the exhibition "Ecology - Ruse 1987." Artists set up the first public notification about the state of the environment. Paintings, photos, etchings and sculptures depicted apocalyptic scenarios. Gas masks, hopelessness, and a poisoned life in grey and yellow were the means to express the situation. This exhibition turned out to be the most visited exhibition Ruse had ever seen. The artistic part was accompanied by some statistics from the local hospitals, reporting the health damage among the population and its dramatic rise during the recent years. Lung disease had increased according to the following figures displayed in Table 1 (G.S., 1988/a):

Table 8-1. People in Ruse with Lung Diseases

| | |
|---|---|
| 1975: | 969/100,000 |
| 1985: | 17,386/100,000 |

Some 3,500 people suffered from an unidentified lung disease for which doctors had created a new medical term, the Ruse Lung. Each worker lost on average 9.5 workdays per year because of some form of lung disease. In the same measure, skin diseases and allergies also multiplied. In 1986, 86,000 children and 62,000 adults (of a population of some 200,000) were considered as outpatients of the medical system. The number of stillborn babies was rumored to be considerably elevated, but official figures did not, and still do not, exist, as they were declared a state secret. During the periods of highest pollution, children were admitted to school only with gas masks. Thousands of families wished to leave Ruse because of the so-called gas attacks. Two thousand of them actually did so, which resulted in a governmental prohibition of moving out. Some residents left clandestinely and lived in the mountains, hidden in shepherds' huts.

The first national alerts to the dramatic situation in Ruse were announced by cultural journals (Baumgartl, 1989). *Narodna Kultura* and *Literaturen Front* called the suffering of Ruse a national shame, and stressed humanist and social arguments. Reports appeared in the Western press in November 1987, namely, *The Independent* and *Le Matin*.

Despite the negotiations between Zivkov and Ceaucescu in October 1987, during which a protocol for removing the causes of pollution in the Giurgiu-Ruse area was signed, the gas clouds continued to plague the population of Ruse. On November 14 and 15, the highest levels of chlorine were registered, exceeding 10 to 12 times the regulations. The next wave arrived during the beginning of February 1988. On February 10, 2,000 people demonstrated in Ruse. As a consequence, *Literaturen Front* published an article with the following passage:

It is painful to live in Ruse. Not only is the environment unhealthy, but also the moral environment suffers, and these unhealthy changes are becoming increasingly serious. The population of this town is suffering from depression, and a feeling of hopelessness prevails. It has lost its confidence in the future.[9]

There were many more comments in this extremely critical style, both in newspapers and the radio. Journalists and readers were touched by the shocking facts and the lack of information over the years. The taboo on Ruse, meant for the media, caused emotional outbursts all over the country. The Bulgarian Writers' Union called the scandal of Ruse "a touchstone of the nation's conscience."[10] In the meantime, the grassroots protest had spread all over the country, with solidarity committees rising in other towns. They were, however, not publicly known until March 1989.

On March 8, 1989, at the Dom na Kinoto in Sofia, after the projection of the film Disaj (Breathe), the Social Committee for Environmental Protection of the town of Ruse was founded. This film tells about the gas attacks over

Ruse and the first protests of the mothers in Ruse during the autumn of 1987. Three hundred sixty persons were the founding members, among them members of the Institute for Philosophy, the Institute for Sociology, the Center for Physics at the Academy of Sciences, the film studio Ekran and the Committee of Young Film-Makers. An administrative council, consisting of 33 members, was elected. It consisted mainly of persons of the arts and culture, professors, poets and film directors. They created a statute, expressing their goals:

- To help and induce the state-run organs to improve the life-sphere in the county of Ruse;
- To pick up the initiative to obtain qualified ecological expertise and to include even foreign experts in order to analyze the ecological situation in Ruse and Giurgiu;
- To consider the risks of earthquakes;
- To convince the Romanian government to dismantle or move the chemical factories of Giurgiu elsewhere;
- To allow full transparency regarding the ecological facts in the region;
- To introduce control by all of the society in the efforts to improve life-quality; and
- To tell the whole truth of what has happened.[11]

The foundation of the Committee was rapidly responded to by the Bulgarian Communist Party (BCP). A few days after the foundation of the Committee, the persecution of its founders started. The government refused to accept the existence of the committee, criticized the way it defamed the authorities, and worried about international tensions with Rumania. Intimidation was followed by observations by and subpoenas from the state security. The members were interrogated and suffered reproofs. The Institute for Philosophy, which seemed to be the basic unit of the uprising, was battered with reconstructions and removals. The productions of the Center for Documentary Films were censored. Some prominent members of the Committee were expelled from the Communist Party and removed from their posts (Ashley, 1989/a).

The party maintained the necessity to improve the environmental situation in Ruse, but rejected activity outside the existing framework, that is, the Patriotic Front. Non-governmental ecological concern was considered as counterrevolutionary activity.

After these directives and repressions by the party, the Committee lost its influence, but its members reappeared later in the Ekoglasnost national movement. By July 1989, all members seem to have joined this new group.

## Ekoglasnost

The Independent Union Ekoglasnost (EG) was founded on March 22, 1989, during a meeting in Sofia. The foundation was officially announced on April 11, 1989, mainly by members of the Ruse Committee. Its chairman, Petar Slabakov, when asked about the relationship between them in July 1989, even called it "an off-shoot of the Ruse Committee" (Engelbrekt, 1989). The founding members wanted to create a legitimate non-governmental organization (NGO) which was also willing to take on political initiatives (Fisher, 1990). Its program was addressed to the general public, and expressed concern about the ecological situation and demanded openness, clarity, and transparency in all policies concerning the environment. EG maintains that the collective management of the economy conceals individual neglect which has flourished in state-owned industry. Hence, pluralism in ownership is essential so that the state can sanction individual owners and managers rather than itself (Ekimova, 1990).

EG paid special attention to environmental damage all over the country. EG also pronounced itself against nuclear energy. They had to struggle for the right of public discussion and to organize as a movement.

The demand for public and democratic control of all environmentally relevant decisions was a radical criticism of core Communist ideology. EG unified the unsatisfied opponents and potential reformers of the Socialist system, among them quite a few members of the Communist Party (Crampton, 1990). To avoid persecution, members or supporters of EG did not have to register formally, but were accepted by personal guarantees of two members. Participation of all interested and affected citizens and groups was solicited and accepted (Fisher, 1990). Its activities, however, were not reported in the government controlled Bulgarian mass media.

From summer 1989 on, EG's target was to gain international attention. It was universally understood that it must present itself as a group solely concerned with ecology, although the members recognized broader political role and goals. When the Conference for Security and Cooperation in Europe held its meeting in Sofia in October 1989, activists of EG seized the opportunity to protest publicly. During probably the biggest non-official demonstration held in Bulgaria since the Communists took power, the ecological right was claimed as a human right. This issue proved to be one of the themes most discussed at the conference. The international recognition and attention transformed EG into a prominent group that was hard to ignore.

Bulgaria had been at the center of international criticism throughout the summer of 1989 due to the exodus of some 300,000 ethnic Turks (Perry, 1991). The government therefore did not want to create another reason to be pressured and internationally blamed (Engelbrekt, 1989). EG tried to remain uninvolved in the ethnic discussion and to present the image of a group of serious ecologists, well-informed on environmental issues.

After the protests in October 1989 and the subsequent removal of Todor Zivkov, who had been the party leader for about 35 years, EG gained a high reputation in Bulgaria. Zivkov's removal was not a direct consequence of EG's activities, however. Rather its protest had the effect of a catalyst for internal fights within the party. The role of EG was recognized even by the new governors when, in November 1989, Petar Mladenov met opposition artists and politicians, among them Petar Beron, the new leader of EG. At the same time, the government tried to weaken the opposition's position by responding tactically to the environmental protest. For example, in November 1989, the Governmental Committee on Environmental Protection was given ministerial rank.

When, the opposition forum Union of Democratic Forces (UDF) was established in December 1989, EG was one of the founding groups, and Petar Beron was assigned the post of chairman. Nevertheless, EG's most vibrant period and its importance or an ecological umbrella organization came to an end at about this time. From that moment, when free political discussion became possible, other issues gained importance. The UDF took over the function of an umbrella for opposition to the Communist Party.

From the beginning, the EG movement had maintained that it would not take part in the government. When, in December 1989, Chairman Petar Beron was offered the post of environment minister in the coalition government of Popov, he refused. EG adapted the position that its members would have to resign from the association to take part in the government.

In his speech at a meeting in February 1990, the leader of Ekoglasnost declared that "Ekoglasnost will always be in opposition to whatever government there is."[12] But this compromise was not enough to calm all disagreements on the course of EG. A two-wing variant similar to a model adopted in Czechoslovakia was frequently mentioned -- one movement focusing on ecological issues and a second wing engaged in party politics, which was prepared to take over governmental responsibility.

By 1990 most members with political ambitions had left EG for the previously founded Green Party (GP). Nevertheless, EG decided to take part in the national elections of June 1990. Three reasons were given (Fischer, 1990). First, if the movement were not politically active in the parliament, it would have no influence in the crucial transition period. Second, the high popularity of EG among the population would increase the votes for the green movement altogether. Third, some members with an interest in political participation had not joined the Green Party. The role of an eternal opposition was now interpreted as taking part only in legislative authority, never in executive power.

### The Green Party

The first public suggestion to found a Green Party was on November 25, 1989. The designated general secretary of Ekoglasnost, Aleksandar Karakacanov, declared "it is high time that Ekoglasnost copes with ecology; whereas, for politics a Green Party should be created".[13] The assembly for the foundation was held on December 28, 1989, at the University of Sofia.

There had been a previous attempt to found such a party during the Zivkov regime. But the Party of the Green Masses, founded by two activists in Vratsa in Northern Bulgaria had only a short life and achieved nothing more than ephemeral publicity in the Western media. Its operators were arrested immediately after the publication of a letter in November 1988; neither were their ideas further elaborated (Ashley, 1989/b).

At the very beginning, the new party was positively commented on even in the press -- which was at that time still rather faithful to the governing Communist Party -- a fact which did not happen to other new political forces. The tactics behind this was to weaken Ekoglasnost. Even sincerely favorable publications assumed that the Green Party (GP) would outstrip Ekoglasnost in membership and reduce the power of this important opposition factor. The brusque manner in which the GP was announced shocked some of the EG's members. Because of this and personal tensions, when, on January 12, the GP joined the UDF, the only vote against its admission came from Ekoglasnost.

The GP spread to other Bulgarian towns, such as Veliko Tarnovo, Ruse and Sliven. In April, four months after its foundation, the GP already existed in 90 places, mainly the bigger towns, and had gathered 10,000 members. The official registration of the GP was accomplished on February 15 by the Sofia city council:

The Green Party of Bulgaria is registered as a political organization. It adopts as its goal the realization of a program in the areas of civil right and political order, social politics, the intellectual sphere, economy and ecology, natural heritage, and interior politics. The aims of the party shall be realized through representation in the National Assembly, the government and local political organs, including all measures allowed by the law.[14]

On March 12, 1990, the first issue of *Ekopolitika*, the weekly newspaper of the GP, was published. It provided a genuine forum for political discussion and ecological issues, but tended to reproduce the style and mode of already existing newspapers. It did not exist for more than a few months.

Soon the GP established contacts with sister parties in other countries. Bulgarian Greens assisted in the founding congress of the Czechoslovakian Green Party in February 1990. In Poland, some of them had the opportunity of participating in a course about election campaigns (Fisher, 1990). Close connections were also established with the Polish Greens; the closest partner in the West was the German Greens. A member of the GP still could attend the conference of the German Greens in early 1990 (Crampton, 1990).

To be able to control ecological damage the GP created its green patrols, which should "carry on a continuous pressure on those who continue to pollute Bulgarian nature."[15] They have no right to fine, but can collect evidence and then present it to society.

By March 1990 there was already preoccupation with the financial situation of the GP. Most of their money had been used for publications and propaganda and on attempts to juridically fight the Communist Party's monopoly on editorials and duplication processes. Not much money was left for an efficient election campaign.

A central issue surrounding the GP was the question of participation in power -- the decision as to whether the Greens should be a party struggling for executive power or, rather, be an opposition party. This was answered with reference to the Western European Green parties. According to the interpretation of the Bulgarian Greens, these "refuse to be limited only to environmental issues, but want to question in a more general framework the way of traditional policy making. However, they do not want to take part in the political scheme of left and right, and try to work out their own political profile."[16] The Bulgarian GP feels very close to this idea, i.e., they do accept power in order to realize their ideas.

In the special situation of the first elections in Bulgaria -- with two political blocks -- they felt constrained to declare themselves. Confronted by the choice between the Bulgarian Socialist Party (BSP) and the UDF, the GP stuck to its membership within the non-Communist coalition. As an opposition to the Communist (left) regime, Bulgarian Greens takes stands which seem to resemble Western right-wing positions. Thus, the first political group in Bulgaria to propose an entry into NATO and the EC was the Greens. They also wanted a Bulgarian military delegation to take part in the Gulf War as a member of the Allied forces. Their economic ideas are clearly related to a free-market economy.

On April 21, the GP held its first national conference. Split up in commissions, they discussed mainly the economic, social, ecological, and intellectual aspects of the party's policies. For the national elections in June, they set up a broad platform. In general, they declared the intention to "seek a compromise between the interests of men, society and nature."[17]

## Party or Movement: A Constant Dilemma

The question of whether a movement should remain as such or convert itself into an institutionalized party is constant in history. It was clearly perceived as such by the members of the workers' movement at the end of the 19th Century. The evolution of this movement to the socialist parties is described by Michels (1911), who concluded that, in all parties, an oligarchy is established. He calls the manifestation of this evolution "the iron law of oligarchy."

In Western Green parties, the dilemma between a party which is prepared to accept political responsibility and power, and a movement which sees itself as a critical opposition to the ruling system, leads to a break in the party. Most famous in this regard is the dispute between the two German wings of Realos and Fundis. Petra Kelley, a leading fundamentalist, even admitted that she is "afraid of the moment when the Greens obtain suddenly 13 percent and, thus, become a party which tries to gain power" (Wiesenthal, 1984).

The Bulgarian Green movement started its political role already split into two organizations which represent exactly these different points of view. The two approaches inside the Bulgarian ecological movement reflect a similar duality on a more abstract level. The two general ways of coping with environmental problems and reconsidering nature are worked out by Lauber (1988):

a) The adversaries to the technical progress give way to the protest movements, and
b) Representatives of the industrial Weltanschauung opt for auto-regulation, which is canalized into state regulation (Lauber, 1988).

As a conclusion, the three stages of environmental protest in Bulgaria follow a chronological development and radicalization of the actors. In a diachronic confrontation, the evolution and penetration of society by the different ecological movements is briefly summarized in Table 2.

Table 8-2. Ecological Groups in Bulgaria: Concerns, Political Role, Extent

| Group | Concern | Power | Extension |
|---|---|---|---|
| Committee | Local | Protest Movement | Regional |
| Ekoglasnost | Strategic | Opposition | National |
| Green Party | Programmatic | Political Party | National |

In general, the potential job-sharing between EG and GP should have been an advantage to avoid tensions.  Nevertheless, tensions did exist, and the relationship between the two of them is still not completely clear.  As just described, these tensions also continue inside EG.

In comparison with Western countries,  Bulgaria has a certain old-fashioned and non-contemporaneous view of ecological problems.  Whereas in Western studies the ecological discussion has reached a highly technical, rational level, the discussion in Bulgaria is still more concentrated on ideological options, strategical questions, and the identity problems of the green activists.

Despite enormous differences between existing economic systems -- at least in Europe -- one common paradigm is opposed by green movements: progress through technical-industrial solutions and economic growth.  Green movements are opposed to systems that emphasize economic and technical progress, whatever their ideological background.  Potentially, their protest against a ruling system seems to resemble the opposition political force.  Actually, their opposition is not due to ideological differences, but to an opposition to any dominating system in Europe.

In the Bulgarian case, this statement can also be read in the opposite direction:  Opposition chooses the green issue to manifest its opposed ideology.  This vehicular function of the environmental issue is obviously limited in time.  The change of the movement's orientation is the shift among the members to more political, rather than purely environmental, aims.

## Review and Outlook

When the Ruse Committee commemorated the second year of its existence on March 8, 1991, the very person who had brought up the idea to found the Committee, stated "We do no longer work for Ruse, neither for ecology; we became egoists."  As far as the environment is concerned, this statement is valid for the greater part of the green movement. All of them have somehow lost contact with the grassroots.

Ruse: As Bulgarian newspapers reported in November 1991, gas attacks on Ruse continue.  Rumors among activists talk of interests from a third (Western?) side, which hinder a definitive solution.  The Committee does not exist any more.  Activists are in one of the factions of Ekoglasnost or the Green Party.

Ekoglasnost: The movement has split up and is in an unusual situation. The Politiceski Klub Ekoglasnost, which is registered as a party, is not in Parliament, whereas the Independent Movement Ekoglasnost within the UDF is part of the governmental coalition.

Green Party: As with Ekoglasnost a split-up occurred, and the main part was voted out of Parliament. The GP overestimated its electorate and, with others, had set up a liberal platform. It gives the impression of a liberal party, but not too much green is left.

The relevance of the green movement could be underlined by the fact that, at the beginning of 1991, the Greens had obtained key positions such as the Sofia mayoralty, the presidency of the UDF, and two ministerial posts in the coalition government of Popov. The vice-president of the UDF parliamentary club in the present parliament is a member of Ekoglasnost, as was the Minister of Finances; and before their parties were reestablished, many leading politicians had joined Ekoglasnost.

The danger for the environment is, however, that in their rush for rapid economic development, all the political forces will forget about the importance of a sustainable development. But the already existing serious environmental problems should remind us that, in all the possible future evolutions or revolutions Eastern Europe might face, the harmonization of development and environment has to be taken in account.

# Notes

1. Due to the scarce number of publications in this field the present paper is based on the following sources: Material collected about Ekoglasnost by Radio Free Europe in Munich, Official publications of Ekoglasnost, Reports and analyses of Bulgarian and Western media, (Rare) comments on Bulgaria in academic literature, and Environmental and political literature on other Eastern countries.

2. Rabotnicesko Delo, 4, 5, 1987.

3. The European Single Act (see Amtsblatt der Europaischen Gemeinschaften, L169, 29. Juni 1989) contains a special section on environmental issues: Viertes Aktionsprogramm der EWG fur den Umweltschutz, KOM (86) 485 endg.

4. Christian Science Monitor, June 11, 1971.

5. Pernik is a curious example of how far the perception of environmental problems and their solutions can influence all scientific domains. Urban plans for whole suburban settlements are defined by the necessity that winds should carry away the polluted air; so houses are constructed in parallel arrangements.

6. Ikonomiceska Misal, April, 1985, p. 55.

7. The multi-layered meaning of this title of a poem is hardly explained by only one translation into English. Apart from the direct "Where will this all bring us to?" also notions of "How far can we reach?" and When shall we not be able to go further? are suggested.

8. Dunavska Pravda, 10, 17, 1984.

9. Literaturen Front, 2, 18, 1988.

10. Narodna Kultura, 2, 12, 1988.

11. Ecopolitika, 3, 12, 1990.

12. Ekoglasnost, 7, 7, 1990.

13. AFP, 11, 29, 1989; Ecopolitika 3, 12, 1990.

14. Ecopolitika, 3, 12, 1990.

15. Duma, 4, 10, 1990.

16. Ecopolitika, 3, 17, 1990.

17. Duma, 4, 22, 1990; Ecopolitika, 4, 23, 1990.

## References

Ashley, S. 1987. Chernobyl and Eastern Europe: One Year After the Accident, Background Report 67 (Eastern Europe), April 24, 1987, pp. 1-4, *Radio Free Europe*, Munich.
Ashley, S. 1988. Politburo Announces "New" Ecological Policy: Bulgarian Situation Report 5, April 27, 1988, pp. 9-13, *Radio Free Europe*, Munich.

Ashley, S. 1989a. Repression of the Independent Human Rights Association Intensifies: Bulgarian Situation Report 1, February 3, 1989, pp. 9-14, *Radio Free Europe*, Munich.

Ashley, S. 1989b. An Attempt to Found a Green Party: Bulgarian Situation Report 1, February 3, 1989, pp. 15-18, *Radio Free Europe*, Munich.

Crampton, R. 1990. The Intelligentsia, the Ecology and the Opposition in Bulgaria. *The World Today*, *46*, No. 2, 1990.

Davy, R. 1990. The Central European Dimension. In Wallace (ed.), *The Dynamics of European Integration*, pp. 141-154.

Ekimova-Melnishka, M. 1990. Ekoglasnost: A Breath of Fresh Air? *Panoscope*, May 18, 1990, pp. 13-14.

Engelbrekt, K. 1989. Bulgaria's Independent Environment: Ekoglasnost: Bulgarian Situation Report 7, August 7, 1989. p. 15-18. *Radio Free Europe*, Munich.

Fisher, D. 1990. Report on a Visit to Bulgaria: Environmental Politics in Bulgaria. Unpublished paper, Ecological Studies Institute, London.

G. S., 1988/a. Air Pollution in Ruse Reaches Alarming Levels: Bulgarian Situation Report 2. February 11, 1988, p. 29-31. *Radio Free Europe*, Munich.

G. S., 1988/b. More Protest Over Air Pollution in Ruse: Bulgarian Situation Report 3, March 8, 1988. *Radio Free Europe*, Munich.

Gavrilov, V. 1990. Environmental Damage Creates Serious Problems for Government: Report on Eastern Europe 21, p. 4-12. *Radio Free Europe*, Munich.

Goldman, M. 1972. Externalities and the Race for Economic Growth in the USSR: Will the Environment Ever Win? *Journal of Political Economy*, *80/2*.

Horvath, A., & Szakolczai, A. 1992. *The Dissolution of Communist Power in Hungary. The Life and Demise of Party Apparatus*. Routledge Editors, Cambridge.

Ionescu, D. 1988. Environmental Disputes with Neighbors Continue: Rumanian Situation Report 5, March 29, 1988. *Radio Free Europe*, Munich.

Jancar, B. 1987. *Environmental Management in the Soviet Union and Yugoslavia: Structure and Regulation in Federal Communist States*. Durham, Duke University Press.

Kaloyanoff, M. 1971. Pollution in East Europe. *East Europe News*, April 1971.

Kelley, D. 1976. Environmental Policy-Making in the USSR: The Role of Industrial and Environmental Interest Groups. *Soviet Studies*, *28*, pp. 570-589.

Kriesi, H.    1990.    *The Political Opportunity Structure of New Social Movements:    Its Impact on Their Development*.    Unpublished paper, presented at the Workshop at the WZB, Berlin.

Lauber, V.    1988.    Zur Politischen Theorie der Naturzerstorung. *Osterreichische Zeitschrift fur Politikwissenschaft, 1*/1988, pp. 79-90.

Oschlies, W.    1987.    *Schwefelstaub auf Rosenbluten. Umweltsorgen in Bulgarien*, Bohlau Verlag: Koln, Wien.

R. T.,    1972.    Environmental Problems in Bulgaria:    Bulgarian Situation Report 5, May 18, 1972. p. 1-5. *Radio Free Europe*, Munich.

Redclift, M.    1989.    Turning Nightmare into Dreams:    The Green Movement in Eastern Europe. *The Ecologist, 19*, No. 5, September/October 1989, pp. 177-183.

Spetter, H.    1988.    Environmental Protection -- An Unsolved Problem in Bulgaria. *Environmental Policy Review, 2*, No. 1, January 1988, pp. 42-45.

Vargha, J.    1990.    *Searching for a Sustainable Future - An Outlook from a Descending Region*.    Unpublished paper, presented at the Conference of the Working Group of Environmental Studies on "Environmental Policy Cooperation between Eastern and Western Countries", European University Institute, Florence.

Weizsacker, R.    1989.    *Erdpolitik-Okologische Realpolitik an der Schwelle zum Jahrtausend*.    Wissenschaftliche Buchgesellschaft: Darmstadt (1991).

Wiesenthal, H.    1984.    *Radikalitat und Rationalitat. Strategieprobleme der Neuen Sozialen Bewegungen und der Grunen*.    Unpublished paper.

Chapter 9

# AN INTERNATIONAL ENVIRONMENTAL CONFLICT ON THE DANUBE: THE GABCIKOVO-NAGYMAROS DAMS

Judit Galambos

Debates over large dams are not rare. In fact, most of the large dams in the world have caused social, economic and political tensions and have become targets of environmentalist criticism.[1] In the 1950s and 1960s dams and hydroelectric plants were regarded as reasonable solutions for energy production and for irrigation. When their harmful environmental and social consequences became known and public, the surge in dam projects has shifted to poorer countries, especially to the Third World, where hundreds of dams are being built even today. But public opposition is growing there as well.

Edward Goldsmith's and Nicholas Hildyard's 1984 book, "The Social and Environmental Effects of Large Dams"[2] extended the critique of dams beyond their ecological effects, challenging the economic rationale, technical design and basic development policies that underlie the decisions to construct large dams.[3] The authors went so far as to conclude that large dam projects have been environmentally, economically, and socially disastrous, and have been a major cause of impoverishment and international debt in the Third World. Their argument convinced many politicians, bankers, development specialists, and even some designers, who have started to see large dams in a more skeptical light.

As we shall see, the history of the Gabcikovo-Nagymaros dam system is very similar to that of many other large dams in the world. We will witness the same kind of conflicts between the same kind of "players in the game", (such as between industry and inhabitants affected, engineers and environmentalists, political power and protesters, etc.) as anywhere else in the world.

However, some interesting lessons might be drawn from the case, since two of the countries involved in the conflict -- i.e. Hungary and Czechoslovakia -- in the meantime underwent major political transformation from Communism to democracy. It provides us with a unique opportunity to observe whether this change has made conflict resolution any easier. Another interesting question that deserves investigation is why is it Hungary which represents environmental interests in her conflict with (Czecho)Slovakia and why the Slovaks are on the pro-dam side of the conflict? Political culture and the level of public environmental awareness should not be so different in two neighboring

176

*A. Vari and P. Tamas (eds.), Environment and Democratic Transition, 176–226.*
© 1993 *Kluwer Academic Publishers. Printed in the Netherlands.*

countries with a shared historical past... Examining the standpoints of the major stakeholders, we shall investigate underlying interests, potential gains and losses as they are perceived, and other conflicts -- such as the ethnic one -- linked to the environmental conflict of the dams. We shall see that the speed and method of political transformation in Hungary and Czechoslovakia also had an impact on the course of events. Last, but not least, the role of Western countries -- especially that of Austria -- in the conflict will be put under scrutiny.[4]

## Project Description

The Danube is the second longest river in Europe, beginning in Germany's Black Forest, passing through Austria before crossing or bordering Czechoslovakia, Hungary, ex-Yugoslavia, Romania, Bulgaria, and Moldavia, emptying into the Black Sea. Austria, which has traditionally relied on runoff from mountain rivers to provide hydroelectric power, built fourteen plants on the Danube. As the river enters Czechoslovakia and Hungary, the river spreads across a vast open floodplain, splitting into three main forks, creating a myriad of islands and wetlands. While the river falls 50 meters per 100 kilometers in Austria, in Hungary its total fall is about 50 meters in the whole Hungarian river section (417 kms), which is clearly not enough for energy production. Therefore the construction of huge reservoirs and a bypass canal was necessary to artificially create a fall, the energy of which can be utilized.

The Gabcikovo-Nagymaros Barrage System (hereafter called GNBS) was planned to consist of three dams and two hydroelectric power plants between river kilometers 1696 and 1862 on the Danube. The Danube was to be dammed at Dunakiliti (on the Hungarian side of the Danube) to flood an area of 60 square kilometers (24 square miles), ending in Bratislava. (From Dunakiliti down to the mouth of the Ipoly the Danube forms the common border of Hungary and Czechoslovakia.) The water level of the reservoir would then rise 6.5 m high above the ground at Dunakiliti. The water was to be routed downstream from the dam to a hydroelectric plant, the second dam and lock at Gabcikovo, via a 17.5 km-long by-pass canal, lined with asphalt and a special plastic (to prevent seepage), running parallel to the Danube 5 km within the borders of Slovakia. The walls of the 300-750 m wide canal were to be built to a height of 18 m (almost 65 feet) above the surrounding gravel and sandy terrain. From that height, the water would have enough momentum to turn the giant turbines of Gabcikovo. The Gabcikovo plant was planned to have 8 turbines and a capacity of 720 megawatts. From the Gabcikovo plant, the water course was to be routed back in an 8 km-long canal to join its old bed at Palcikovo/Szap.  The whole canal was to be on Slovakian territory and international water traffic was planned to be diverted to the canal from the 31 km-long section of the "old" river-bed, where only 2.5 percent of the present average discharge would flow, except during inundation after heavy rainfalls.

From the mouth of the canal 100 km downstream, the Danube bends southward, entering a steeper valley by Nagymaros in Hungary.  This was planned to be the site for another hydroelectric plant, the third dam, with the capacity of 158 megawatts.  The Nagymaros plant was planned to work continuously, but producing only two-seventh of the total energy produced by the whole system.  The Gabcikovo plant was intended to produce peak power in the mornings and evenings.  During most of the day the water of the river would have been retained in the reservoir at Dunakiliti, in order to be released twice a day, flushing the reservoir and canal like a toilet, surging through the turbines at Gabcikovo.  This would have in turn created great changes in the water level.  The Nagymaros dam was to be built in order to compensate for these changes (Figure 9-1).

According to the Czechoslovakian-Hungarian bilateral agreement, the two countries were to share the construction work and costs equally, and, after completion, also the electricity produced (3,775 million kilowatt-hours per year).[5]

# History

Plans for utilization of the hydroelectric power of the Danube already existed before the first world war. However, planning of a joint Hungarian-Czechoslovakian investment only commenced in the early 1950s. The plans of the dam system were approved by the Council for Mutual Economic Assistance in 1961.  The governments of the two countries agreed to develop a joint investment project in 1963, the plans of which were completed by 1973. In May 1976 the two countries concluded an agreement about a Joint Agreed Plan on the complex utilization of the Danube for energy production, navigation, and for the management of the water supplies.  This goal was to be reached through the construction of the Gabcikovo-Nagymaros Barrage System. The prime secretaries of the communist parties of the two countries -- Kadar and Husak -- announced the decision in September 1977.  On September 16, the bilateral treaty was signed by the two prime ministers.  The ratification documents were exchanged in June 1978.

Czechoslovakia started construction in April 1978, two months before official ratification.  The Hungarians were less enthusiastic:  shortly after work began on the Hungarian side, public debates over the project began, first in professional associations.  In November 1981, an article harshly

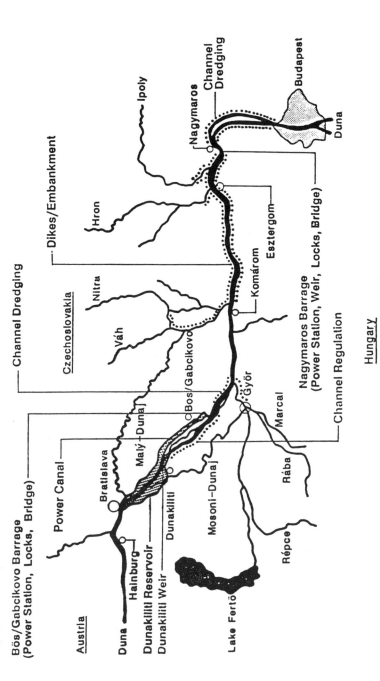

Locus Plan
Áttekintő Térkép

criticizing the project was published by a biologist, Janos Vargha, who later became a leading figure of the environmental movement in Hungary.[6] Czechoslovakia resented that the publication of such an article was allowed in Hungary.[7]

The nervousness of the Czechoslovakian government was understandable. Two month earlier, the two countries agreed to suspend construction work, because of lack of necessary financing. The Hungarian government unilaterally decided to postpone all work until 1990, and initiated a study on the ecological consequences of the dam system. However, in the several expert committees that were formed, dam engineers managed to assert their point of view.

The Hungarian state and party leaders were more concerned about the lack of investment capital than about ecological consequences, therefore they proposed that Czechoslovakia should build the whole project alone -- in exchange Hungary would pay off half of the investment costs with electric energy. The Hungarians also approached Austria. In November 1983, Chancellor Sinowatz and the Hungarian government began to negotiate an Austrian loan and participation of Austrian companies in the construction work.

The Hungarians did not manage to "escape" from the project -- Czechoslovakia only agreed to take over some of the work. In October 1983 the prime ministers of the two countries signed a modification of the 1977 treaty, according to which the completion of the project was postponed by five years. The Hungarian Politburo had already made a secret decision in favor of project completion in June.

In December of 1983 the Hungarian Academy of Sciences completed a report, according to which construction should not be continued until an environmental impact assessment is prepared. In the spring of 1984 public debates were held in university clubs and professional associations. The first grass-root environmental group in Hungary, the Danube Committee, was established in January 1984. The movement collected more than 10,000 signatures in support of a petition, addressed to the Parliament and government, demanding a halt to the construction. The movement grew in size but was not structured. It was therefore sought, unsuccessfully, to found an official association. However, the political leadership toughened its position, prohibiting public discussion and publications against the GNBS.

Finding itself unable to be registered as an association, the movement founded the unofficial Danube Circle. (It only became a registered organization in 1988.) The Danube Circle broke the ban on public discussion of the GNBS by publishing the "News of the Danube Circle" in samizdat. The bulletin contained documents of debates, information on the historical and political background of the project, and an account of the debate in Austria on the Hainburg hydroelectric plant. In December 1985 the Danube Circle received the Right Livelihood Award (the so-called Alternative Nobel Prize). It wished to use the prize money for scientific research into the ecology of the Danube region by creating an endowment for this purpose, but it proved impossible to realize this under the political conditions of the time. This, again, caused a crisis in the movement.

While the Danube Circle was paralysed, three other movements appeared for a short period: one gathered signatures, demanding a referendum; another, the "Blues", demanded that the Parliament should discuss the case and decide on it; the third one, the "Friends of the Danube", demanded that at least the construction of the dam at Nagymaros should be stopped. In January 1986, a letter with 2,500 signatures, protesting the project and calling for a referendum, was submitted to the Hungarian Presidential Council (a body which exercised the functions of the head of state.)

Negotiations between Hungary and Austria for a credit agreement were underway. In January 1986 the Danube Circle, together with Austrian and German environmentalists, held a press conference, protesting against Austrian financing of the project. The Danube Circle also sent a petition to the Austrian Parliament. In February a "Danube Walk" was organized by the Danube Circle and the Austrian Greens, which was violently disrupted by the Hungarian police. The government's action was internationally condemned and the European Parliament passed a protest resolution. In April prominent Hungarian intellectuals published an advertisement in an Austrian daily, Die Presse, asking the Austrians to protest against their government's involvement in the GNBS. In May 1988 there was a demonstration against Austria's role in the case in front of the Austrian embassy in Budapest.

However, the agreement between Austria and Hungary was signed in May 1986. Austrian banks were to supply loans for the construction of the project, and Austrian companies were to be given 70 percent of all building contracts; Hungary was to repay the loans by delivering electric energy to Austria, from 1996. Two-thirds of Hungary's share of electricity produced by GNBS was to be paid to Austria over a period of 20 years, mainly during the winter months, when the level and the flow of the Danube is at its lowest, therefore the GNBS alone could not have provided the required amount of electricity, and new Hungarian power stations would have had to be built in order to amortize the energy debt. The Austrian companies began construction at Nagymaros in August 1988.

Starting from 1988, the political climate in Hungary changed and opposition was revitalized. Because the work at Gabcikovo was already advanced, public attention was focused on Nagymaros and peak-time production of electricity. In August 1988, 15 environmental groups came together in the Nagymaros Action Committee. In September 1988 a two-day international conference was organized by the World Wildlife Fund (WWF) and the Danube Circle on the environmental impacts of the project. Austrian environmental groups, the Austrian Minister for the Environment, representatives of the Hungarian Academy of Sciences, the International Rivers Network, and the Hungarian Union of Young Democrats were also among the participants.

There were further demonstrations and petitions before the October 1988 session of the Hungarian Parliament. This Parliament was different from the one a few years before. While the one-party system was preserved, for the first time in communist Hungary, at the 1985 elections voters were allowed to choose between more than one candidates for one seat in the Parliament. A small number of independent politicians and even some representatives of the opposition succeeded in becoming elected. Thirty-two independent parliament representatives initiated a re-examination of the project and proposed a referendum in the summer of 1988. The proposed debate in the Parliament took place in October 1988, after the first mass demonstration (30,000 people) against the project. However, the representatives -- under strong pressure from the government and the party leadership -- voted for the continuation of the project. The communist party treated voting as a loyalty test: it obliged its members in Parliament to reject the referendum. The formation of a parliamentary party fraction in a one-party system raised serious questions as to the legitimacy of the Parliament and also the possibility of preserving the one-party system at all.

The decision was followed by an even more heated public debate. On October 30, an international day of protest was held, coordinating simultaneous demonstrations against the GNBS in 27 capitals of the world, organized by a Hungarian American, Bela Liptak and his Foundation to Protect the Hungarian Environment. There was a political struggle also within the party leadership. The reformist wing of the party was gradually gaining ground, and in November 1988, Miklos Nemeth, a prominent reformist, became Prime Minister. The party reformers needed popular support in their fight against the conservative wing of the party (and were therefore more receptive to the claims of public opinion and the opposition) and mainly used the GNBS case as a political tool in their struggle.

However, the new Prime Minister sent a delegation to Czechoslovakia in February 1989 to sign a protocol concerning the speeding up of the work. (It was a step difficult to explain. Probably the reformists were not strong enough yet to implement sharp change in the government's policy.) But three months later, in May 1989, his government unilaterally suspended work at Nagymaros for two months. Probably Nemeth wanted to avoid the referendum, demanded by environmental groups. (They collected more than 140,000 signatures in support of their demand by April 1989.)

The government asked the Parliament to authorize a new round of negotiations with Czechoslovakia on a modification of the 1977 bilateral agreement and to commission expert bodies which would examine the consequences and possible decision alternatives.[8] In June, the Parliament - - the same representatives who had voted for continuation in October -- approved the government's proposal.

The still orthodox communist Czechoslovakia accused the Hungarian government of submitting to political pressure by the opposition, and labelled the decision anti-socialist and hostile to Czechoslovakia, jeopardizing the "good neighborly relations" of the two countries. The government of Czechoslovakia demanded completion of the project in its original form, and threatened with a claim for compensation of at least 2,700 million Czech crowns.

The Austrian government declared that it would respect the Hungarian government's decision, but it also had to defend the interests of the Austrian contractors, therefore it hoped to reach an agreement with Hungary on due compensation. The Austrian claim for compensation was about 2,000 million Austrian schillings in addition to the bill of 850 million for the work already done by the Austrian contractor.

In May 1989, after the Hungarian government suspended construction, a goodwill delegation of Hungarian representatives and environmentalists travelled to Vienna to ask the Austrians not to demand too much compensation from the Hungarian government, because this would cause further economic difficulties for Hungary. They also sought to convince their partners of the importance of preserving the natural values of the Danube and the area around it. The delegation met the representatives of four parliamentary parties, the constructing company Donaukraftwerke, the Verbundgesellschaft, the Lander Bank, the Kreditanstallt, and the Austrian green movements. The Austrian Freiheitspartei and the Green Party showed sympathy for the Hungarian standpoint. The Greens were of the opinion that the Austrian contractors should limit their claims to the costs of work already performed and should renounce any further compensation; Austria should not stick to the energy transport contract, because it will force the Hungarians to build a new power station. However, the discussion between the delegation and the representatives of Donaukraftwerke ended in mutual disappointment.

In July 1989 the Austrian economic minister declared that the Austrian contractors claimed "only" 2,600 million schillings, provided Hungary would pay the sum immediately in cash. In case Hungary chose to amortize its debt by supplying energy, this sum would be increased by the due interests. The minister rejected the Hungarian proposal that the Austrian firms could be compensated by new contracts (e.g., road-building) in Hungary. The Hungarians considered the claim for interest unfair, since Hungary did not denounce the energy-supply treaty. The contract with the Austrian companies terminated in November 1989.

In July Nemeth met with Adamec, Prime Minister of Czechoslovakia. Nemeth announced a prolonged suspension of construction at Nagymaros till October 31, and that work would also be suspended at the Dunakiliti dam, thus preventing the diversion of the Danube into the bypass canal, built on the Slovakian side of the Danube. (Without the Dunakiliti reservoir, the power station at Gabcikovo could not be operated.) The Hungarian Prime Minister offered alternatives for joint revision -- suspension for 1, 3, or 5 years, or total abandonment. (These were later rejected by Czechoslovakia in a diplomatic note.) They decided to nominate a joint expert committee to examine the case thoroughly and to discuss the disputed facts and arguments. They also agreed to extend the suspension period until the end of October, when they would make the final decision together. However, the joint expert committees were unable to produce anything meaningful. According to the Czechoslovakian experts, no new facts could be identified in the period after 1977, that could justify the modification of the original contract.

In August 1989, in a letter to the Hungarian Prime Minister, the Prime Minister of Czechoslovakia gave notice of a "possible provisional solution", in case the Hungarians suspend construction for a longer time or completely abandon it. Czechoslovakian politicians and experts spoke of "new technical solutions" which would make the unilateral diversion of the Danube possible: thus they would not need the reservoir at Dunakiliti for operating the power plant at Gabcikovo. The Hungarians suspected that Czechoslovakia was only bluffing, trying to put pressure on the Hungarian decision-makers in order to make them comply with the original treaty.

In September, technical and legal experts of the two countries met in order to discuss the controversial issues. The meeting ended without success. On September 22, the Hungarian government decided that the Nagymaros dam would not be built up. The Hungarian Prime Minister informed the government of Czechoslovakia about the decision in a letter. On October 11, the two prime ministers met to discuss the issue. No success again.

On October 31, the Hungarian Parliament took a stand on abandoning the Nagymaros power station and dam. The Parliament approved the government's report on the investigations of expert bodies into the economic, ecological and technical aspects of the project. Another resolution of the Parliament in November authorized the government to initiate a modification of the 1977 inter-state agreement. The government was obliged to act on the basis of ecological considerations and the priority of national interests. The conclusion of a new bilateral agreement on ecological guarantees was regarded necessary before the Gabcikovo plant would be put into operation. On November 30, the Hungarian Ministry of Foreign Affairs submitted an official proposal for modification of the 1977 treaty. According to the proposal, the dam at Nagymaros would not be built, and the power plant at Gabcikovo would not be operated before satisfactory ecological guarantees were elaborated. The government of Czechoslovakia did not give an official response to this proposal.

This compromise proposal of the Hungarian government was resented by the environmental movement because -- in exchange for the abandonment of the Nagymaros part of the project -- it would have allowed Czechoslovakia to complete and use the Gabcikovo dam, provided it gave appropriate ecological guarantees. According to the environmentalists, such guarantees were meaningless if the Gabcikovo plant was operated, since its construction and operation in any form would have a very negative effect on the environment.

In the Fall of 1989, radical political changes in Czechoslovakia gave the Hungarians high hopes for a solution to the conflict. The relations of the two countries had become very tense because of the growing difference in their political systems. The orthodox communist Czechoslovakian leadership had regarded the reform communist Hungary as a dangerous example for its people, therefore it had reacted nervously to Hungarian moves. The Hungarian side hoped, that Czechoslovakia would "catch up" during its "Velvet Revolution", and it would be easier for two democratic regimes to work out a solution to the problem. Therefore, the Hungarian premier made a new proposal in January 1990, in a letter addressed to the new Prime Minister of Czechoslovakia: he suggested that the final decision should be left to the new governments, to be formed after the first free elections in both countries. The environmental organizations also demanded that no work should be carried out on the site, except maintenance, until free elections were held in both countries. The Czechoslovakian Prime Minister rejected the suggestion.

The new Czechoslovakian leadership used a dramatically different, more positive tone than its predecessor. The environmentalist groups in Czechoslovakia were no longer denied the freedom of speech and freedom of the press, and their activities were not prohibited any more. As a result, environmentalists in Czechoslovakia could activate the public more effectively against the project. In February 1990, the first major protest rally against the Gabcikovo project took place in Slovakia, launched by a Slovakian group, Eurochain. More than 60,000 people took part, forming a chain between Hainburg (Austria) and Komarom (Hungary), along the Danube. In April 1990, environmentalists of Austria, Czechoslovakia and Hungary adopted a Danube Charter, in which they called for the establishment of a trilateral nature reserve to protect the Danube landscape. They also demanded that the Gabcikovo constructions be halted.

The new government of Czechoslovakia desisted from the unilateral diversion of the Danube and agreed to set up expert committees to investigate the ecological and economic aspects of the project. In February, Vladimir Lokvenc, the commissary of the investment, who had fanatically defended the scheme for decades, was dismissed by the Slovakian government. The new political leadership was more receptive to ecological concerns than the previous government and readily admitted that the idea of the whole investment had been a mistake. However, it still insisted on completing construction because of the huge sums of money already invested, and because construction work was allegedly 90 percent complete.

Slovakian decision-makers were convinced that the hydroelectric plant at Gabcikovo, after having acquired the necessary secondary installations to protect the environment, should produce energy in order to amortize the investment costs. However, they were prepared to accept the less risky, continuous mode of operation, instead of the originally planned peak-time energy generation, which would have been more disastrous to the environment. Hardliners, who even wanted to put the question of Nagymaros and peak-time energy generation back on the agenda, started to re-gain influence in the Slovakian government. Supporters and opponents of the project were fighting each other before the elections in Czechoslovakia.

Seven alternatives for the future of the dam system were developed by the dam builders. "Variant A": completion of the project according to the original agreement of 1977. "Variant B": completion of Gabcikovo according to the original plans, but abandonment of Nagymaros. The Gabcikovo power station would only work in continuous mode, but not producing peak power. "Variant C": completion of the Gabcikovo hydroelectric station with a smaller reservoir, exclusively on Slovakian territory. The Danube's diversion would start a bit more upstream, where both banks of the Danube are in Slovakian territory. Here a new dam would be built near Cunovo. Nagymaros would be abandoned. "Variant D": completion of the Gabcikovo plant, but without the reservoir in Kortvelyes and without Nagymaros. "Variant E": a limited use of the Gabcikovo plant for energy production, shipping, and discharge of floods. "Variant F": suspension of all construction works, conservation of the installations. "Variant G": limited use of some of the installments for flood protection, and gradual dismantling of the others; the landscape would be restored to an ecologically acceptable extent. These alternatives were never officially submitted for discussion to Hungary.

The environmental groups rejected all of the above alternatives, because all aim at some -- at least partial -- operation of the installments, and they regarded the environmental risks of each version unacceptable. They developed their own version, "Variant H", according to which the by-pass canal would be demolished, the Danube would flow in its original river bed, nature would be restored, and the dams at Dunakiliti and Gabcikovo would serve recreation purposes.

The Czechoslovakian Ministry for Forestry and Water Management set up two commissions, which completed their work by February 1990. The first committee, which studied the interrelationship of the dam system and the quality and quantity of the surface and underground waters, pointed at the huge environmental risks, due to sedimentation, and leaking of toxic sediments into the underground waters of Csallokoz, in case of filling the reservoir at Kortvelyes. According to the committee's opinion, the condition for filling the reservoir would have been a basic improvement of Danube's water quality. However, there are no sewage treatment facilities in Bratislava, and there exist no plans for preventing seepage of oil from the nearby petroleum refinery either. The investigations of the second committee, which focused its work on the other environmental impacts of the dam system, concluded that a reduced operation of the hydroelectric plant was possible on the following conditions: 1. The flow of 1,000-1,500 cubic meters must be guaranteed in the original river bed; 2. The functional relationship between the Danube and its lateral arms must be maintained; 3. The natural dynamics of the river must be preserved. These conditions were never met by the dam-builders.

The first free elections took place in the early spring of 1990 in Hungary and in early summer in Czechoslovakia. In Hungary a mid-right coalition led by the Hungarian Democratic Forum formed the government. In Czechoslovakia the liberal Civil Forum won the elections.

The new Hungarian government, formed after the first free elections, announced its political program in May 1990. It disapproved the completion of work at Gabcikovo, and promised rehabilitation of the landscape at Nagymaros as soon as possible. It also announced its intention to start negotiations with Czechoslovakia on sharing the damages. The political program's details, related to the GNBS, were handed over to the Czechoslovakian commissioner on May 31.

The new Hungarian government made a quick deal with Austria, although paid a high price for it. Donaukraftwerke made its claim officially in May 1990: 3,850 million Austrian schillings, including 250-300 million schillings as interests. According to Donau Kraftwerke (DOKW), the cost of work already done was 950 million, and the rest consisted of materials already ordered, the cost of preparatory work, unrealized profits, etc. This sum would be amortized by supplying electricity. After several months of bargaining, a compromise agreement was reached in November 1990. According to the agreement, Hungary had to pay 2,6 billion Austrian Schillings (US$ 240 million) as compensation for expenses and lost profit.

After the elections in Czechoslovakia, negotiations between the two countries were resumed. In August 1990, the foreign ministers met in Prague, but produced no results. In September, the Federal Ministry for Forestry and Water Management set up expert committees for the assessment of the seven variants for the future of the GNBS. When the various committees completed their work, a so-called synthesizing committee, nominated by the Economic Council of the Federal Government, should have summarized their findings and drawn conclusions. However, this committee was not established. Instead, Minister Oberhauser recommended "Variant C" to the Slovakian government, based on arguments of the economic committee, but disregarding the opinion of the other committees, and also of the Water Management Institute, which did not recommend "Variant C". The Environmental Committee of the Slovakian Parliament did not oppose "Variant C" as a temporary solution, provided its 19 conditions were met, among them ecological ones.

The negotiations between the two countries were at a stalemate throughout 1991 and 1992, although there were several exchanges of letters and meetings of their politicians. The internal tension within Czechoslovakia was growing, and the separation of the federal state into two nation-states became a reality. The GNBS issue was more and more seen by Czech politicians as a Slovakian issue, and they were less and less willing to interfere, although formally the representatives of the federation were Hungary's partners during the negotiations. In the meantime, construction at Gabcikovo steadily continued.

On April 16, 1991 the Hungarian Parliament authorized the government to negotiate with Czechoslovakia on the termination of the 1977 treaty and to initiate the conclusion of a new treaty, which would address the consequences of the cancellation of the project. (It also meant that the government was not authorized to negotiate solutions other than the termination of the treaty.) A week later the Parliament adopted another resolution, suggesting the conclusion of an ecologically oriented, comprehensive Danube treaty with Czechoslovakia. The resolution also called for the establishment of transboundary national parks on both sides of the Danube, in Austria, Hungary and Czechoslovakia. With this resolution, the Hungarian Parliament basically approved the suggestions of the environmentalists.

In 1991 the focal point of environmental protest was in Slovakia, especially in Csallokoz (Wheat Island), the region which was most affected by the Gabcikovo dams. In February 1991, 82 mayors of Csallokoz addressed a petition to the governments in Prague and Bratislava, and to President Havel, requesting the project's termination. In July, environmental activists of Eurochain occupied a pump station at the construction, in order to prevent the planned filling of the 25 km-long canal of the power station. (The purpose of this operation was to stop the further melting of the asphalt sealer of the canal, due to hot weather.) The protest action was supported by other environmental groups, like the Slovak Union of Nature Protectors, the Czech Human, the Austrian Global 2000 and WWF-Austria, the Hungarian Danube Circle and Reflex. The occupation lasted for more than a month, and government representatives and representatives negotiated with the protesters. At other occasions, however, they were brutally beaten, arrested and removed by the police. The prefilling of the canal began on July 27, in spite of the protest.

International environmental groups condemned the project again and again. Protest against Gabcikovo was part of the "International Coalition Against Large Dams" campaign, organized by environmental groups in protest of the World Congress of the International Coalition On Large Dams (ICOLD), held in Vienna on June 16-17. Eurochain, Global 2000, the International Rivers Network, the Austrian chapters of Friends of the Earth and World Wildlife Fund, the Austrian Ecology Institute and the Austrian Students Union sent an open letter to ICOLD, reminding dam builder engineers of their responsibility for the public good.

In July 1991, eight organizations of WWF adopted a resolution addressed to the government of Slovakia, in which they demanded the suspension of construction works and a detailed environmental impact assessment study. In August, public debates over the project went on, with the participation of Slovakian academic institutions as well as environmentalists. In September, an international petition, initiated by WWF Austria, the International Rivers Network, Friends of the Earth, and the Slovak Union of Nature Protectors, addressed to President Havel and Slovakian Prime Minister Carnogursky, demanded the immediate halt of construction at Gabcikovo.

On July 15, 1991 negotiations were held between the two countries, with the participation of the Slovakian Prime Minister, Jan Carnogursky. Czechoslovakia confirmed its intention to complete the Gabcikovo plant; in case the two sides were unable to agree, Czechoslovakia would implement it in a go-it-alone effort -- "Variant C" -- diverting the Danube onto Slovakian territory. The Hungarian delegation threatened with sanctions if this occurred. The setting up of a joint commission of inquiry was considered. Czechoslovakia suggested that experts of the European

Communities be included in the commission, which should specify a technical solution for the operation of the Gabcikovo plant. Instead, Hungary proposed a bilateral committee for the assessment of ecological risks.

On July 23, 1991, the Slovakian Government decided to implement "Variant C" as a "provisional solution" (i.e. until Hungary agrees to complete the project according to the original plans). On July 25, the Federal Government approved the Slovakian decision. Work began soon thereafter. In August the Hungarian Prime Minister sent a letter to his Czechoslovakian counterpart, expressing his hope that the conflict can be solved by patient dialogue.

Federal politicians felt they were in a difficult situation: caught between two -- Slovakian and Hungarian -- intransigent positions. They were convinced, that a compromise could be reached, if the Hungarian government's hands were not tied by the resolution of the Hungarian Parliament. Certain members of the Hungarian delegation told privately their Czechoslovakian colleagues, that they favored the completion of the Gabcikovo plant.[9] Therefore, Federal Minister for the Environment, Josef Vavrousek requested for a hearing by the Environmental Committee of the Hungarian Parliament. The hearing was held on September 11, 1991. Vavrousek tried to convince the representatives to extend the mandate given by the Parliament to the Hungarian government, so that it could also investigate solutions other than the total abandonment of the project. He argued that the Slovakian Parliament supported "Variant C", which was a technically viable solution, although a difficult one and very harmful for the environment. If the Hungarians were not willing to accept the method of negotiation suggested by him, the federal government would not be able to prevent the implementation of the unilateral Slovakian solution. He suggested that a trilateral commission be set up, with the involvement of EC experts, in order to investigate all realistic proposals and their potential environmental impacts. The Hungarian representatives did not want to change the mandate of the government, but they were ready to examine other proposals as well. They regarded the involvement of EC experts desirable, however, not giving them the role of an arbitrator.

The issue of the joint expert committees was further discussed on December 2, at an intergovernmental meeting in Budapest. The two delegations agreed that it would be reasonable to set up a joint expert committee, with the participation of EC experts, for the revision of the whole issue. The Hungarians insisted that the committee's activity only had sense if the construction of "Variant C" was not continued. The Czechoslovakian delegation declared that even a temporary suspension of the construction was impossible. The above standpoint of Czechoslovakia was officially reinforced on December 18, in a letter addressed to the head of the Hungarian delegation.  One day later, the Hungarian Prime Minister sent a letter to his Czechoslovakian counterpart, in which he expressed his concern that the setting up of the planned joint committee would not be possible.  The Czechoslovakian Prime Minister in his reply of January 23, 1992 repeated his government's position.

The next letter came from the Hungarian Prime Minister on February 26, in which he confirmed that Hungary was ready to accept the establishment of a joint expert committee, but only if construction was suspended by Czechoslovakia.  Otherwise Hungary would be compelled to withdraw from the 1977 treaty.  On March 24, 1992, the Hungarian Parliament adopted a resolution on the unilateral termination of the 1977 treaty by Hungary, in case Czechoslovakia would not suspend construction of "Variant C" before April 30.  On April 1, Hungarian Minister Ferenc Madl met with Jan Carnogursky, Slovakian Prime Minister, and explained to him, that a trilateral committee could elaborate an objective proposal for a solution, based on expert opinions, during a six-month suspension of construction works.  Carnogursky did not trust that the Hungarians would at the end accept the opinion of the trilateral committee.  On April 23, 1992 the Prime Minister of Czechoslovakia, in his reply to the Hungarian Prime Minister's letter of February 26, called the Hungarian request an ultimatum, and announced that the construction of "Variant C" would not be suspended, instead, it would be completed by October 31, 1992.  As a result, a verbal note on the unilateral termination of the 1977 treaty by Hungary (taking effect from May 25) was handed over to the Czechoslovakian Embassy on May 19, 1992.

For a long time, many Hungarians did not believe that the construction of "Variant C" should be taken seriously.  Both the government and the environmental groups, especially Janos Vargha and the Danube Circle, were convinced that this option was not viable from a technical and financial point of view, and it was only used by the Slovaks for blackmailing the Hungarians, in order to make them finally return to the original treaty.  However, the determination of the Slovakian government to proceed with the implementation of "Variant C", in spite of all technical and financial difficulties, was greatly underestimated.

In a press conference held in April, financier Hannes Androsch -- former Vice Chancellor of Austria -- announced that of the US$ 350 million necessary for the completion of Gabcikovo, approximately US$ 150 million would be raised by him from sources outside the country by June.  The Androsch International Management Consulting was hired by the Slovak Office for Hydraulic Engineering.  Earlier Androsch had been director general of the (Austrian Credit Institute) and responsible for the funding of Nagymaros.  Immediately following Mr. Androsch's announcement,

Austrian government officials distanced both themselves, and the Austrian government from the finance plans. Austrian Chancellor Vranitsky and Foreign Minister Mock announced that the Austrian government would not guarantee Austrian loans given to the project. In a letter to Austrian Chancellor Vranitsky, Hungarian Premier Antall emphasized that loans emanating from Austrian financiers would provoke severe political repercussions and do serious damage to relations between Hungary and Austria. Androsch approached both the World Bank and European Bank for Reconstruction and Development (EBRD), but neither of them provided the requested credits. Thierry Baudon, Director of the Infrastructure, Energy and Environment Department of EBRD declared that EBRD would definitely not be involved in a contentious project of dubious economic value and negative environmental impact which was argued between two of their member countries.

Other Austrian companies were also involved in the construction of Gabcikovo: Keller-Grundbau, which constructed the latest part of the canal, and three Austrian gravel companies -- Korzil- Baukeramik, Porr Limited AG, Wiener Lieferbeton GmbH, which founded a joint venture, the Donau Schotter GmbH (Danube Gravel Ltd. -- with Hydrostav, the Slovakian constructor, to sell 2 million tons of gravel per year from Gabcikovo.

International and Austrian environmentalists protested against the involvement of Austrian companies in the construction. Major U.S. NGOs[10], such as the Audubon Society and the Sierra Club, announced a boycott against Austrian ski-industry and tourism. WWF-Austria, Global 2000, and Ecoropa urged the Austrian government to immediately stop Austrian involvement in the construction and guarantee that no energy produced in Gabcikovo would be imported by Austria. They wanted the Austrian government to support the Hungarian policy in the issue, and to become an active mediator. They were also concerned with ground water in Austria, in the region not far from the dams (north-eastern Burgenland and Lower Austria), therefore they expected the government to intensify research on groundwater movements in the region. They also demanded the enforcement of a law requiring environmental impact assessment for activities of Austrian companies in foreign countries.

After the June 1992 elections a new government was formed in Czechoslovakia. In Slovakia the rather nationalist and separatist Democratic Movement for Slovakia formed a government, headed by Vladimir Meciar. The program of the new Slovakian government stressed that both Gabcikovo and the nuclear power station at Mochovce were necessary for Slovakian independence.

In July 1992 the new Slovakian government reviewed the status of the Gabcikovo investment. Experts prepared background materials for both the Federal and the Slovakian government, examining several alternatives, including temporary suspension of Variant C, due to financial difficulties. However, before the experts could complete their examinations, the Slovakian government already indicated that it would not support a solution involving a suspension of the construction. It was decided that financial problems had to be solved by August 15. Three billion kcs. (Czech Crowns) were needed for 1992, and five billion for the following two years. The government had only 1.4 billion kcs. for 1992 at its disposal. Until 1992, altogether 15-16 billion kcs. were spent on the investment. Although the dam was allegedly 90 percent complete, the revealed figures did not reinforce this statement.

The government considered several solutions for raising the missing amount of capital, such as credits from the West, privatization of certain facilities of the dam system (e.g., the planned recreation areas), Western investors, etc. In September the government issued bonds in the value of approximately $1.25 million for financing Gabcikovo. At the same time it launched an international propaganda campaign, in order to improve the image of the project. As part of the propaganda, Slovakian premier Meciar, during his visit to Austria at the end of October, offered that the Bohunice nuclear power station, inconveniently close to the Austrian border, would be closed down, if the Gabcikovo hydroelectric plan was put into operation. This offer was interpreted as blackmailing by the Austrian press.[11]

In October the Hungarian government decided to provide financial resources for the rehabilitation works in Nagymaros. Work would start in the spring of 1993.

The Slovakian government planned the diversion of water from the Danube to the new canal for October 20. An alliance of environmental groups, the Danube Defense Action Group[12] organized a series of public events between October 16-25, protesting against Gabcikovo, ending with a peace march on the site. WWF, Global 2000, and Ecoropa organized parallel actions. Bela Liptak organized a peace march at the construction site. In Bratislava an international press conference was held by WWF, where WWF's legal study on "Variant C" was presented. However, the diversion of the Danube could not be prevented.

Since bilateral negotiations have not produced any result for the last few years, the idea of a third party involvement in conflict resolution has repeatedly come up. In June 1990, the environment ministers of the EC decided in Dublin to offer funding for an international expert study of Gabcikovo. The Slovaks also proposed the Hungarians to jointly ask the EC that an investigation on possibilities for protection of groundwater in the region of the dam be supported by the Poland-Hungary Assistance for

Restructuring the Economy (PHARE) program. The Hungarians, who insisted on the abandonment of the whole project, and had the results of their own investigations on the risk of the dam system on underground water, declined the offer.

However, Hungary agreed to request -- jointly with Czechoslovakia -- an impartial expert assessment from Bechtel Environment Inc. (USA) and Hydro Quebec International (Canada).[13] The results of their investigations have been interpreted differently by the dam-builders and the environmentalists. According to the dam-builders, both firms agreed that it was possible to deal with the expected ecological consequences. However, Peter Drossler, representative of Global 2000 said at a press conference,[14] that the specialists of Hydro Quebec recommended in their report a detailed study of environmental and social concerns regarding the project, especially about impacts on groundwater, security of infrastructure, and impacts on the fauna and the people. They recommended a number of changes and corrections to the infrastructure. However, none of these recommendations were followed by Hydrostav. In a letter of October 16, Hydro Quebec declared its intention not to become involved in the project because of its concerns about it -- emphasized Drossler.

After being rejected by Hungary, Czechoslovakia applied for support from PHARE for a solution to the problem of groundwater. The proposal was accepted, and the problem is being studied in cooperation with Danish and Dutch companies. However, according to environmentalists, this research program cannot provide results before 1994; up till now only provisional computer models exist on the groundwater around Gabcikovo. The results of this study might come too late from the point of view of saving the groundwater, since the most significant interference into the water system happened in October 1992, when the Danube was diverted from its original river-bed.

In a letter addressed to the foreign ministers of the two countries in April, the Vice President of the Commission of the European Communities expressed the readiness of the Commission to take part in the resolution of the dispute, provided that both countries refrained from steps that could influence or anticipate the future conclusion of a (Slovak-Hungarian-EC) trilateral committee. The Hungarians interpreted this condition in a way that did not allow Slovakia to continue the construction of "Variant C". The Slovaks maintained a position according to which the 'temporary' solution did not constitute an irreversible step.

In September the two prime ministers, Antall and Meciar, met and agreed about the establishment of a trilateral fact-finding committee, which was to prepare trilateral discussions. The committee started its work in October in Bratislava, and by October 31 completed a report on the status of the Gabcikovo dam, specifying which works could be suspended and under what conditions, and which works had to be continued for safety, navigation, and flood control reasons. The report was kept secret.

The Danube Circle was suspicious of the trilateral fact finding committee, and questioned the neutrality of the EC expert participating in it, because he had worked with the Slovakian member of the committee in a PHARE project. The Danube Circle was also dissatisfied with the composition of EC experts in the trilateral committee, on the ground that most of them worked for the hydraulic engineering industry, and there was no expert of the environmental impact of hydroelectric plants among the committee members. They were not pleased with Hungary's representation in the expert committee either. The Danube Circle found it alarming that the report was kept secret as well as the negotiations.

On October 28, trilateral negotiations were held in London, where a non-binding understanding was signed, spelling out a working schedule towards a negotiated solution. Czechoslovakia was to charge 95 percent of the Danube's water to the original river bed (this solution technically excluded energy production, but navigation was to take place in the bypass canal), and stop the turbines of the hydroelectric plant. The understanding foresaw bringing the dispute jointly before a binding international arbitration body, or the International Court of Justice in the Hague. On the basis of the fact-finding committee's report, another trilateral committee was to be established, mandated to determine by November 15 which parts of the hydroelectric project should be terminated, continued or suspended. A week before the London agreement the Hungarian government wanted to turn unilaterally to the International Court in the Hague, and laid a complaint to the emergency conflict resolution institution of the Conference on Security and Cooperation in Europe. After the London agreement Hungary withdrew its complaint.

The paragraph of the London agreement which required the Slovaks to charge 95 percent of the water into the original river-bed, was technically impossible to meet. Even if the canal was only used for navigation, more than five percent of the water would flow into the canal, during the shiplocks are in use. It was also very unlikely that the Slovaks would not want to use the hydroelectric plant. The Slovaks were supposed to stop the turbines in Gabcikovo and construction of Variant C by November 21, based on the recommendations of the secret fact-finding report. The Slovaks simply disregarded it, and went on with testing the turbines. The federal government's attempt to enforce the London agreement and stop construction resulted in a federal cabinet deadlocked in a 5-5 vote along Czech and Slovak lines. By early November, the diversion of the Danube was declared completed by the Czech and Slovak authorities. Hungary's Foreign Minister declared that the river's diversion and the completion of

the related structures violated Hungary's sovereignty and territorial
integrity, breaching international law, as well as the terms of the London
understanding.    Slovakia referred to the work completed so far as
temporary and reversible, and was ready to suspend temporarily work on
certain ancillary facilities according to the recommendations of the
trilateral committee. (Most of these works were scheduled for 1993 or
1994, anyway.)

Due to construction mistakes, the Slovaks were unable to open the lock
of the completed dam, therefore they could not discharge the promised
water into the old river bed. As a result, the branches of the Danube dried
up. A couple of weeks later a Slovak barge carrying concrete blocks sank in
the artificial canal in front of the lock that releases water back into the
natural branch of the Danube. The barge caused the locks to malfunction
and blocked the flow of water to Hungary, further intensifying the
environmental damage. Because of the environmental emergency situation,
the Hungarian Parliament decided to give 100 million Hungarian Forints
for steps relieving damage.   A Hungarian ad hoc committee of local
governments in the region, together with local water management and
nature conservation experts, proposed that the reservoir at Dunakiliti
should be put in operation, in case the trilateral negotiations do not lead to
the abandonment of Variant C. Environmentalists and members of the
Environmental Committee of the Parliament opposed this solution,
because they felt it was an unacceptable compromise, threatening with
Hungarian "retreat" at Dunakiliti, thus encouraging the Slovaks. The water
management authorities developed solutions the Hungarians could
implement, in order to alleviate environmental damage.  They argued that
it was urgent to prevent a threatening environmental catastrophe, causing
irreversible harm, which would happen within 100 days, if the region did not
get the promised amount of water.   According to the most optimistic
variant, the old river bed would get the promised 95 percent of the water,
which would save the branches and the groundwater. Although in principle
it was possible, the installments at Gabcikovo were not adequate to charge
more than 30-50 percent of the natural flow.  Even in this case some
technical and water management measures could alleviate the damage. For
the worst case, when only the amount of water arrived, that was originally
planned in Variant C, the construction of a temporary sill would become
necessary at Dunakiliti, so that the water in the old river bed would be
raised, and the branches would get some water. The reservoir at Dunakiliti
would help in regulating the water.

Hungary officially complained about Slovak non-compliance with the
London agreement, concerning the amount of water charged into the old
river bed.  In its answer the Czechoslovakian Ministry of Foreign Affairs
declared that Czechoslovakia complied with the London agreement,

however it did not document it. In the meantime, international legal experts of the two countries negotiated in Prague about the text of the joint petition to be made to the International Court in the Hague. However, they were unable to agree. While Slovakia wanted to bring the case of Hungary's unilateral termination of the 1977 treaty and Czechoslovakia's claim for compensation to the Court, Hungary intended to ask for a decision on the Slovakian breach of contract, i.e. the unilateral diversion of the Danube.

The next round of trilateral negotiations was held in Brussels, on November 28-29, 1992. The trilateral expert team presented its finding at the meeting. The two countries agreed, that the case would be jointly taken to the International Court of Justice in the Hague. The text of the joint petition remained to be decided by the two countries. While the final decision is expected from the International Court, which will definitely take some time, the two countries need some temporary solutions for the in-between period. The Hungarians wanted to get guarantees that 95 percent of the flow would be charged to the old river-bed, according to the London understanding. It seems no agreement was reached in this question. The results of the negotiations were interpreted so differently by the Slovaks and the Hungarians, that it is difficult to know what in fact happened in Brussels. The Slovaks felt they won and were authorized to continue the construction. The Hungarian delegation emphasized the importance of the agreement on the joint petition to be made to the Hague. However, no agreement was reached in its text and the two sides were unable to reach a satisfactory compromise on the question of temporary solutions. This is where the case stood at the time of writing this study.[15]

## Stakeholder Analysis

### Dam Builders

The water management bureaucracies, planners, hydraulic engineers and the construction companies belong to this category. Under communism, where state ownership was predominant, the industry was closely intertwined with state bureaucracy. It was also true for the water management sector, which became very powerful in both Hungary and Czechoslovakia.

In the years after World War II, even prominent water management specialists emphasized the serious technical, economic and ecological limitations to utilizing water power in Hungary. But in the early 1950s, the gigantic projects of the Soviet Union, such as the hydroelectric plants at Stalingrad and Khuibishev on the river Volga, set the examples to be followed. However, this was not just a phenomenon of the communist societies: there was a general push to construct dams around the world, partly motivated by the illusion of success promoted in the United States and the Soviet Union.

Large dams are key instruments of a development ideology that relies on a political approach of top-down, centralized economic planning. In many countries, dams are lobbied for by powerful interests seeking to use the central government to achieve political goals, such as industrializing a country's economy, or subsidize certain economic sectors. These political goals often override objective economics: in order to keep the project going, capital costs and risks are frequently underestimated; decommissioning and dam-failure costs are ignored, while benefits are calculated assuming a best-case scenario for operation. There is a general failure to recognize that the negative physical, social and ecologic impacts of a project can be far greater than any planned benefits. It is true that some sectors of the population benefit from the dam projects, but others bear the costs.[16]

When it came to dam building, the situation in Hungary and Czechoslovakia was not different than anywhere else in the world, except that the above was even more true. In the communist economy, the various ministries and state-owned companies competed with each other for the limited resources available from the state budget. Bankruptcy was impossible in a planned economy: partly for ideological reasons, and partly because bankruptcy of a state-owned company would have meant that the management, the controlling bureaucracies or the planners had made serious mistakes. Since nobody wanted to admit such mistakes, the company was always granted extra resources from the state to ensure its survival. Economic considerations played no role, on the contrary, the firm which showed the worst financial balance was rewarded. This was also true for large investments: once the construction had been started, it was easy to get more and more resources to complete the project. Therefore the costs and difficulties were usually underestimated considerably at the outset, in order to influence decision-makers in favor of the project. If new difficulties arose in the course of realization, the same bureaucracy removed them in order to ensure employment for a growing number of people, especially bureaucrats, for a long time.[17]

The Hungarian and Czechoslovak water management bureaucracy had been planning the building of dams on the Danube for flood control and navigation purposes since the beginning of the century. However, it was a very costly, infrastructural investment, what was considered "non-productive" in communist economies. In order to obtain the necessary support and resources, planners and the water management establishment started to emphasize the energy production side of the project. Since for the same expenditure a thermal power station could have been built with a capacity twice as great as that of the hydroelectric plant, energy production from the planned hydroelectric plant could not be competitive due to the unfavorable natural conditions, the costs of the investment had to be curbed through manipulations. For example, the investment costs were reduced by the amount of money provided by the energy sector and other sectors were charged with the expense of the necessary additional investments (e.g., sewage treatment).

The planners and the supporters of the project have been seeking to justify the project with the following arguments:

- The presumed energy needs of the Hungarian and the Slovakian economy could not be met without the energy produced by the hydroelectric stations of the GNBS.

- The dam would protect the Danube lowlands from floods.

- The project would result in improved navigability on a section of the Danube which is otherwise the most difficult to navigate, and where the necessary navigation depth could not be reached without damming. With the dams the 2.5 meter navigation depth recommended by the Danube Commission can be reached, and the two countries would be connected to Europe through the Danube and the Danube-Rhine-Main Canal.

- The project would promote the irrigation of agricultural lands above the dams.

- The GNBS would produce electricity at peak consumption times, replacing the air-polluting brown coal operated power-stations in Slovakia. (Recently it sometimes allegedly replaces a nuclear power plant as well.)
- Hydroelectricity is a renewable source of energy.

- Concerning ecological impacts, it was claimed that they could be mitigated. Moreover, dam builders argued that those risks were not proved, therefore they did not have to be taken into account. More recently, it has even been argued that the project would have positive environmental impacts, and it is necessary for saving the Danube and the ecological balance of the region, which suffers from a diminishing water supply, due to past mismanagement of the river.

- The dams offer recreation and sport opportunities.

- Since the eighties it has often been argued that abandonment of the project was impossible because the money already invested would be lost. This is still the main argument in Slovakia, which has almost completed construction at Gabcikovo. The Hungarian dam proponents argued that further high costs would be caused by restoration and the payment of compensation to Austria and Czechoslovakia, not to mention the political and economic harm caused to Hungary by not fulfilling its international legal obligations.

The argumentation of the participating Austrian companies was similar. Hannes Androsch defended the project at an energy conference organized by the Financial Times in London, in November 1992. He argued that Gabcikovo would change the ratio between nuclear energy (60 percent of total) and hydroelectric energy (only six percent) in Slovakia. The project would also have a positive effect on flood protection, navigability of the Danube, drinking water supply and protection of ground water. He said that concerns about low ground water existed for several years and seismic sensitivity was neither proved nor disproved. According to Androsch (and water management specialists) the ecological problems of the Danube did not originate in Gabcikovo; however, $80 million would be devoted for their solution. The investment costs of Gabcikovo were so far approximately $1 billion plus 130-170 million for environmental expenses, but the estimated yearly income from energy production would be around $120 million.

Many dam builders honestly believe in these arguments. Although dismantling of the project could provide employment and economic benefits to the dam builders to the same extent as its construction did, it is more gratifying for engineers to build a dam than to dismantle one, and also offers better career opportunities. The water management sector remained relatively intact and powerful even after the political changes in Hungary and Czechoslovakia. Their opinions about the project do not differ too much in the two countries. Sometimes members of Hungarian expert delegations expressed privately that they agreed with their Slovak colleagues. The Slovakian press often used arguments of Hungarian water management engineers to demonstrate the benefits of the dam system, and to prove that the nature of the Hungarian decision to terminate the project was political, not scientific.[18]

Concerning the Austrian companies participating in the construction of Nagymaros and later Gabcikovo, they were looking for new opportunities after the construction of the Hainburg hydroelectric plant had been stopped by popular protest in 1984. The well-established Austrian dam-building industry, facing a decreasing selection of new sites and growing public opposition at home, became a major dam-builder abroad, especially in the Third World and in Eastern Europe, where massive repression and human rights violations have been connected with projects like large dams.[19] Several controversial hydro-power projects have been built with the contribution of Austrian money and technology all over the world.[20] Dam-builders had to face less obstacles in countries where public protest was illegal, decision-making was done amid secrecy, and economic and ecological considerations were overrun by political ones.

## Public Opposition

The Hungarian, Czech and Slovak, Austrian and international environmental groups, ecologists, biologists and other scientists, part of the local population and their local governments belong to this category. Occasionally they were supported by politicians -- usually those in opposition. They have been concerned with the environmental and safety risks of the project, and the huge human, social, and economic costs involved. However, the most upsetting factor for most people was that the history of the project was characterized by non-democratic decision-making, secrecy, and suppression of public opinion. The main arguments of the project opponents were the following:

- The dam system would jeopardize the fresh water supplies of millions of people both in Hungary and Slovakia, through the complete loss of possibilities for water production by bank-filtration in the area, and through the contamination of subterranean water reserves in the area, the largest untapped underground fresh water reserve in Europe.

- The dams would destroy valuable natural flora and fauna by upsetting the ecological balance of the river and radically changing the level of the subterranean water.

- The agriculture, forestry, and fishing of the region -- the economic prerequisite of the local population -- would also suffer serious losses.

- The construction of the GNBS would result in the disappearance or dramatic transformation of a historic and natural landscape and ruin valuable archeological remains.

- Geological and seismological considerations were neglected in the plans, although huge reservoirs often cause earthquakes.

- Flood control could be solved by cheaper and safer means and, above all, in the case of an accident the flooding engendered would be more extensive than in the past; yet no measures were taken to prevent such an incident or protect the surroundings. The dam system was not safe due to poor construction practices observed.

- Navigation conditions could be improved by much cheaper and simpler means.

- If a more efficient use of energy was implemented, the energy produced by the GNBS would not be needed.

- The amount of energy produced is insignificant when compared to the enormous ecological harm it causes.

- The project is an economic nonsense, since Hungary will have to hand over two-thirds of its share in the produced energy to amortize its debts to Austria for 20 years, which is normally the life-span of such dams.

- All the stated benefits of the project could have been realized by safer and cheaper means, without raising credits, without diverting resources from more important goals and without ruining other -- more valuable -- resources.

- The project would trap a few villages of Slovakia -- in which the population is predominantly Hungarian -- between the Danube and the artificial canal, isolating them from the surrounding settlements.

When planning of the dam system started, environmental concerns were not among the priority issues to be taken into account. It was the same all over the world then. In many countries environmentalism became a political force in the late seventies and in the 1980s. In Austria, an already completed nuclear power station could not be put in use due to public protest in the late seventies. A hydroelectric plant could not be built up at Hainburg for the same reason in 1984. A joint Austrian-Czechoslovakian plan to construct a hydroelectric plant at Wolfsthal could likewise not be realized. Environmental groups have become quite powerful in Austria by the eighties. They enjoyed all the democratic freedoms, therefore protection of the environment could remain their sole concern.

Austrian environmental groups have closely cooperated with the Hungarian, Slovak and Czech environmentalists. They not only assisted them in their fight against the project, but tried to prevent Austrian involvement in the construction and financing of both the Nagymaros and the Gabcikovo plants. They sent petitions to Austrian representatives, published leaflets and expert studies, held international press conferences, participated in demonstrations. In 1988, as a sign of protest, Global 2000 occupied the building of the management of Donaukraftwerke. The Austrian Green Party submitted protests to the Austrian government. In 1986, some representatives of the Austrian Green Party distributed leaflets in Budapest against a visit to Nagymaros by Austrian officials, this leading to several detentions and expulsions. Some representatives of the Austrian People's Party used the case as a tool in their political fight with the Socialist Party of Austria, therefore supported the environmentalists. Austrian environmentalists did a lot to prevent Androsch from raising credits required to complete the unilateral construction of the Gabcikovo plant according to "Variant C". The Austrian government had to officially distance itself from Androsch's activity in order to avoid open conflict with Hungary, but also embarrassment at home.

Austrian groups, being more experienced than their Hungarian and Slovak colleagues, rendered them invaluable assistance in calling international public attention to the case, and in putting them into contact with international environmental groups. The Austrian chapter of WWF was especially helpful in this. They were even able to produce expert opinions by respected international experts. WWF helped in the preparation of an expert study, which substantiated the Hungarian government's decision to stop construction at Nagymaros.[21] In June 1990, WWF also offered the Czechoslovakian government to provide consultation and render a detailed expert study regarding the Gabcikovo power plant. WWF published four expert studies, analyzing various aspects of the case: "Statement of WWF to the Gabcikovo-Nagymaros Project", "Energy for Slovakia: Options for Environment-oriented Policy", "Goods Traffic on the Danube", and "Variant C of Gabcikovo-Nagymaros under International Law".

In Hungary, the case triggered the evolvement of the environmental movement. Since all kind of political opposition was prohibited and suppressed in communist Hungary, this environmental protest, within a short time, assumed a political character. In order to achieve their environmental goals, Hungarian environmentalists had to fight for democratic freedoms as well. For this reason, they enjoyed the support of the political opposition. Open attack of the political system was not tolerated by the regime. However, it should also be taken into account, that the Hungarian regime, compared to, e.g., East Germany or Czechoslovakia, was a "soft dictatorship", where the opposition faced censorship and loosing their jobs, but were relatively rarely beaten by the police or arrested. Since about the mid-1960s, the political leadership was divided between conservative communists and those who wanted to introduce economic, and even political reforms. During those periods, in which the reformists were stronger within the leadership, political debates were better tolerated. The political fight within the communist party became more and more open in the second half of the eighties. Since the party and the government itself was divided on the issue of the dam system, the case was used as a political tool by the reform communists, especially after the removal of Kadar, First Secretary of the Communist Party.

Initially the environmental movement tried to convince those politicians who were not ardent supporters of the investment. The environmentalists used all possible legitimate methods, such as petitions, request for registration etc. Since official registration was not granted to the movement, it had to function underground. This limited the number of participants in the movement, which therefore could not present itself as a membership organization; it had no political program or political tactics.[22] Since information on the dam system and the environmental movement was censored, the interested public could only receive information on these issues from the samizdat publications of the Danube Circle.

From the mid-eighties the political power tried to respond to the grass-root initiatives by organizing environmental organizations centrally, top-down, with the intention to replace the grass-root initiatives with these "official" organizations. At the same time, demonstrations organized by the Danube Circle between 1986 and 1988, where Austrian environmentalists also participated, were brutally dispersed by the police.

The 1988 parliamentary session, where representatives voted for the continuation of the construction under strong political pressure of the communist party, was the first parliamentary session in the history of communist Hungary, which was televised from the beginning to end: masses of voters witnessed the manipulations in the Parliament and the behavior of their representatives. After the parliamentary debate and voting, in several constituencies citizens -- supported by the political opposition -- started to collect signatures in order to recall those representatives who had spoken in favor of and had voted for the continuation of construction at Nagymaros. This initiative grew into a campaign for getting rid of orthodox communist representatives.

The GNBS investment became a symbol of the old regime, against which all opposition forces were fighting. The reformist wing of the Communist Party did not want to follow the unpopular policy of the old regime concerning the Gabcikovo-Nagymaros dam system. The new government of Miklos Nemeth, after suspending construction, drew the environmentalists, at least formally, into the work of expert committees, preparing reports for the government and the Parliament on the future of the investment.

After the ruling communist party dissolved itself, most members of those movements which participated in the Nagymaros Committee, joined the new, emerging political parties. Those who remained in the Danube movement, took part in the exhausting and frustrating work of the committees. However, it was expected that the new parties will pay more attention to the environment, since the environmental movement played an important role in the dismantling of the old regime. Although the new parties promised major changes in the country's environmental policy, the environment was not a major issue at the first free elections. The new political parties were busy with creating their ideological image, and the environment did not seem to be the most suitable issue for this purpose.[23]

After the first free elections, the Danube movement did not function as a movement for a long time. Since all the major political parties and the new government committed themselves to the decision of not building up the Nagymaros dam, public pressure did not seem necessary any more. Although there were disagreements between environmentalists and government politicians on the tactics to follow in negotiations with Czechoslovakia, there was a general optimism -- especially after the political changes in Czechoslovakia -- that the case could be solved by negotiations. The environmentalists focused their efforts on lobbying in the Parliament, preparing expert materials and suggestions, participating in committees, putting pressure on and negotiating with government officials.

Dissatisfaction with the government's slow and hesitating policy on this issue led to the re-establishment of the Danube Circle in 1991. However, it did not take the Slovak intention of implementing "Variant C" seriously. Danube Circle, under the influence of Janos Vargha, held the opinion that the Slovak government announced this variant with the intention of blackmailing the Hungarian government, in the hope that the Hungarians would agree to complete the project according to the original plan, rather than letting Slovakia unilaterally divert the river. They were convinced, that Slovakia did not have the financial and technological means for implementing "Variant C", therefore pressed the government not to negotiate about anything else than termination of the treaty. Mobilization of the public started too late, when the construction was almost complete, in spite of desperate signals from the Slovakian environmentalists concerning the seriousness of the Slovak government's intentions and the quick pace of its implementation. Creative suggestions for alternative negotiation tactics were not developed either.

In Czechoslovakia the communist regime was more repressive than the Hungarian one. There were some differences between the two republics of the country. In the Czech Republic opposition could only appear underground, for example in the framework of Charter 77, and criticism could only be published in samizdat. In May 1989, an individual wrote an open letter to federal Prime Minister Adamec, calling for changes to the Gabcikovo system and a stop to construction at Nagymaros. The letter had collected over 3,600 signatures, several times more than any other such petition. It was followed by an increased administrative harassment (arrests, confiscation of passports etc.) of the unofficial environmentalist circles.

While opposition in the Czech Republic was not tolerated, in Slovakia criticism within registered organizations could be more extensive. The first professionally valid -- although not comprehensive -- criticism of the project came in 1976, from two institutes of the Slovak Academy of Sciences. The Ecological Section of the Academy produced the first full critique of the scheme in 1988, which was then published in January 1989.

The most active opposition came from the Bratislava city branch of the Slovak Union of Nature and Landscape Protectors (SZOPK). SZOPK published its first proclamation on the GNBS in its magazine in 1981. Soon thereafter, the magazine was suppressed for six months, and censorship in general was tightened up for a while. Nevertheless, in 1988 the magazine was able to devote a whole issue to the case without penalties. In 1987 SZOPK published a proposal for a "Danube River Landscape National Park", this being extended in 1988 to the idea of an International Park. Hungarian and Austrian environmentalists endorsed the Danube Proclamation (which was the only open joint Czech-Slovak venture), including the proposal for the International Park.

After the democratic revolution of December 1989, harassment of environmentalists came to an end, and there were high hopes for a change in the country's policy on the issue. However, those politicians who had opposed the scheme while in opposition, were more cautious after coming to power. The investment became a national symbol in Slovakia, and federal politicians were reluctant to further strain the federation already shaken by Slovak separatism. Nationalist politicians in Slovakia treated the case as a political conflict between Hungary and Slovakia, and not as an environmental conflict. Therefore those environmentalist groups in Slovakia, which continued their fight against Gabcikovo, were labelled as "traitors" and "Hungarian agents" by Slovak nationalist politicians and by the press.

Political change in Czechoslovakia was more sudden, not as gradual as in Hungary. The dam system did not become a symbol of the old regime in the eyes of the public as in Hungary, and opposition against it did not play a special role in the change of the political system. On the contrary: democracy provided an opportunity for Slovak independence, and

Gabcikovo has been depicted as an important tool of Slovak independence, by reducing the dependence of the country on energy imports. It has also been celebrated as an achievement of Slovak technology and engineers, as a symbol of the development of Slovakia. The Slovakian press has remained dominated by those who support the project, and opposers' possibilities to publish their counter-arguments and mobilize the public have been curbed. SZOPK and Eurochain, despite their courageous actions, in spite of international support from environmentalist, and support by the local population affected by the project, have been unable to convince and mobilize the majority of the population of Slovakia. Unfortunately, the protest of the local population and their municipalities in the vicinity of Gabcikovo did not weigh so much with the Slovakians, since the majority of inhabitants of this area belong to the Hungarian minority, therefore their views were considered as biased in favor of Hungary.

## The Governments

### Czechoslovakia

It might be difficult for non-Central-European readers to understand why and how hydroelectic plants could become objects of national pride and ethnic hatred. For this, it is important to understand the symbolic value of the Danube in the history of both nations. Most Hungarians are convinced that the following historic facts have to do with present Slovak behavior.[24] To what an extent is it true? It is difficult to judge. However, perceptions are not less important than "real facts" in conflicts like this one.

In the nineteenth century certain leading Czech nationalists sought to create a united Czech-Slovakian state, whose borders would reach the Danube, which would connect their country with the sea and with the other Slav nations. The possibility of realizing this plan arose at the Paris Peace Conference, at the end of World War I, where it was settled that the southern border of the new state -- created from Bohemia and Slovakia, which had earlier belonged to the Austro-Hungarian Monarchy -- would extend to the Danube, and thereby incorporate ethnic Hungarian territories into Czechoslovakia. The Czechs and Slovaks had originally wanted to get large Hungarian territories also on the other side of the Danube, but the Peace Conference did not approve this.

In the political situation that prevailed during the inter-war period, Czechoslovakian politicians were afraid that they would not be able to keep the ethnic Hungarian territories and the Danube. Therefore the country's authorities did their utmost to mix the Hungarian population along the Danube with Slovaks. They continued this policy after World War II as well: Hungary was pressed by the victorious great powers to come to an agreement with Czechoslovakia on a "mutual exchange of population". This meant that Prague was allowed to expel as many Hungarians from Slovakia as to match the number of Slovaks moving from Hungary to Czechoslovakia. However, the "exchange of population" failed to effect the de-Magyarization of the riverside to the extent planned, since the number of Hungarians was much greater in Slovakia than the number of Slovaks in Hungary. Some people in Hungary feel that the disruption of this ethnic bloc will be completed by the diversion canal of the GNBS.

If we consider that Slovak-Hungarian tensions are rooted in history (Hungarian domination of Slovakia before World War I; incorporation of Hungarian territories by Czechoslovakia after World War I; hostility between the two countries between the two world wars) and that both countries seem to return to the unfortunate historic pattern after more than 40 years of forced friendship, the following factors are also taken into account from the point of view of national interests of both countries: the canal and the dams will turn Bratislava (the capital of Slovakia) into a big international port, taking over the present role of the Csepel Free Port in Budapest. The importance of such a port is underlined by the opening of the Rhine-Maine-Danube canal, connecting Slovakia with Western Europe, and making Bratislava an important port of East-West navigation. There are plans to extend the broad gauge railway of ex-Soviet Ukraine up to Bratislava, which would open up further opportunities for Slovakia as a future transfer point of East-West trade.

Also, by constructing the diversion canal, Czechoslovakia could completely obtain a 31 km section of the Danube. While the original plan would at least have given direction over the water supply of the original river bed to Hungary through the Dunakiliti dam, the unilateral "Variant C" gives this power into Slovakian hands. If Slovakia is not willing to give the promised water to Hungary, the Hungarians cannot do anything against it. This could become a strategic weapon in the hands of Slovakian nationalists, if relations between the two countries remain unfriendly.

It is interesting to note, how small the difference is between various political regimes when it comes to the demonstration of the power of the regime: hydroelectric plants and dams, just like skyscrapers, are modern equivalents of pyramids, cathedrals, and triumphal arches.

It is also true that Slovakia has invested much more money in the construction than Hungary. The argument that withdrawal from the project was impossible because of the high amount of money already spent on it worked even in Hungary for a while. It worked even better in Slovakia. Although Slovakian ecologists, biologists, and water management specialists pointed to the even higher losses that would be caused by the operation of the dam system in the water reserves, nature and agriculture, decision-makers disregarded these opinions.  It can partly be explained by political reasons, as explained above, and partly by the fact that there was a strong personal interpenetration of groups interested in the construction and decision-makers, even after the fall of communism.  In the past regime, directors of Hydroconsult (the company constructing Gabcikovo) were in high government positions.   After the political changes this tendency continued:  the director of Hydrostav, Ivan Carnogursky, was at the same time vice-president of the Slovak Parliament.  Decisions were mostly made by technical experts, engineers, who disregarded the "non-professional" opinions of environmental and water management experts.

While Czech politicians would have been more compromising, they could not and did not want to act in the federal government against the will of the Slovak government.  While it was important for the Czechs to maintain good relationship with Hungary, it was equally or even more important for them to preserve good relations with Slovakia, whether within a federal state or in two separate states.  To a certain extent, this conflict might have contributed to the increasing acceptance of Slovak independence by the Czech Republic, which thus could get rid of the economic and political burdens of Gabcikovo.

*Hungary*

When the dam system was planned, environmental concerns were not taken as seriously as today.  Even if we take this fact into account, it is clear that the construction of the dam system was more a Slovakian interests than a Hungarian one.  Why did Hungarian decision makers then approve the plan in 1977?

It is true that "transformation of nature" was an important element of communist ideology, which was not restrained by the "capitalist" categories of efficiency and profitability.  However, this ideology was the strongest in the 1950s, when Hungary built up a huge steel industry without the necessary raw materials, and produced cotton in spite of unfavorable natural conditions for cotton-growing.  But even the party leadership of the 1950s rejected the plan of a diversion canal on Slovakian territory.  Why in the 1970s was this decision made, after those capitalist categories were re-introduced into the Hungarian economic thinking during the reform period launched in 1968?  Since the decision was made amid secrecy by the top political leadership, we can only guess.

One of the explanations could be that the decision was made in a period when conservative communists regained power in the political leadership and the reforms were halted.  At the same time, the oil crisis of 1973 raised questions concerning energy supply, which strengthened efforts for self-supply. The Hungarian water management and engineering sector became quite powerful by this time, and was able to convince decision-makers about the need for hydroelectricity. Since such decisions were typically made in secret, the idea was not discussed by a broad circle of professionals, therefore decision-makers were not "bothered" by counter-arguments. Those who were interested in this secrecy, probably found convincing enough reasons: whether foreign policy interests (Soviet shipping interests), or energy policy considerations, or something else, we do not know.

Even in the issue of the diversion canal, the Hungarian leadership was very compliant. While the peace treaties after the two wars had determined the middle line of the navigational route on a certain section of the Danube as the state borders between Czechoslovakia and Hungary, the by-pass canal was to divert this route, thus changing the conditions under which the borders had been set. It had to be decided whether the border should be moved to the middle line of the new navigational route i.e. the by-pass canal. According to the joint investment plan of 1973, the question of borders was to be settled in a separate treaty. However, in 1973 the Czechoslovakian side proposed "the re-examination of the necessity and expedience of the proposed modification of the border". After 3-4 years of high-level negotiations, the Hungarian party leadership agreed with Czechoslovakia that the border would not be moved to the new navigational route and approved the 31 km-diversion of the Danube's water into Slovakia.

As a result of the second oil crisis of 1979-80, a second wave of reforms started in Hungary. This is when professional debates started about the investment, and some departments within the state administration (especially the energy sector) started to attack the scheme on economic grounds. Such battles were kept within the administration, but secrecy was not preserved completely. Hungary initiated the suspension of the construction, and the government commissioned the Hungarian Academy of Sciences and the National Office for Environmental Protection and Nature Conservation (NOEPNC) to review and reassess the investment plans. However, the fight within the political leadership continued, and by the time the studies were completed, the top party leadership had already decided to continue the project. Since neither the report of the Academy nor that of the NOEPNC suggested the continuation of the construction, the Academy's report was kept confidential, the director of the NOEPNC was dismissed, and the office was disbanded in the following year. After the financing of the project was solved by Austrian involvement, no open discussion of the investment was allowed. The conservative wing of the party overcame again.

However, after Gorbachev came to power in the Soviet Union, political fights in the party leadership started again, and in 1988 Kadar was removed from party leadership.    However, the new leadership was no less conservative, although it tried to pretend that it was following the Gorbachev line.   A new Ministry of Environment Protection and Water Management was established, the leadership of which was among the most stubborn supporters of the dam system.   As it was described in the historical chapter, the anti-GNBS movement became quite strong then, and the Parliament discussed the issue.   The scandalous, manipulated voting discredited both the Parliament and the regime.

The reform-communist Nemeth government implemented the sharpest change in Hungarian government policy in the history of the project, by unilaterally suspending the construction.   As the last government of the regime, it was more sensitive to public pressure than its predecessors or its democratically elected successor.  However, it did not follow up its decision by determined actions, e.g. calling the responsible persons to account, or starting to dismantle the Nagymaros dam.   The Ministry for Environment and Water Management disagreed with the government's decision, and some kind of construction went on as "maintenance works", thus raising the bill to be paid to the Austrian contractors.   Irreversible steps were not taken, although the contract with the Austrian company was terminated. Various expert committees prepared reports, but no course of action for the future was set.   The Nemeth government was similarly indecisive in its diplomacy toward Czechoslovakia.

The new democratic government inherited the problem from its communists predecessors.  Since the project was discredited so much, and because of the role it played in the change of the political system, it would have been impossible for the new politicians to continue the joint construction with Czechoslovakia -- most of them committed themselves against the project while in opposition.  However, the new government also inherited the almost intact dam-builder lobby.  The new government was inexperienced and had to rely on the experts of the old bureaucracy, including those of the dam-builder lobby.

The Parliament's declaration of intention to terminate the treaty with Czechoslovakia was made after a long period of political fights.  Two years passed since the decision to suspend construction.  One more year passed until the government terminated the treaty.  Even then it was reformulated as a condition, and it took two more months for the government to declare the termination.  This hesitating behavior encouraged the Slovakian dam builders to continue and even speed up construction.   Private but unconcealed sympathies of certain Hungarian negotiators towards dam building and completion of the project also undermined the official policy.

On the other hand, no constructive suggestions for solution were developed.

The history of the bilateral, and later trilateral negotiations shows that on the one hand, the Hungarian government was uncompromising, but on the other hand, it made no creative suggestions to solve the conflict and tolerated all the violations of agreements by Slovakia. It is true, the Hungarian government was not in an easy situation, especially because certain nationalist politicians in Slovakia adroitly linked the conflict over the dam to the fate of the Hungarian minority in Slovakia. Curbing of their minority rights overlapped with accusations questioning their loyalty to the Slovak nation and state, but testing this loyalty by which side they took in the dam conflict. The Hungarian government might have felt that it had to be yielding if it did not want to harm Hungarians living in Slovakia. On the other hand the Hungarian government did not have similar tools in its hand to coerce Slovakia, and under the conditions of deep economic crisis in Hungary, the government could not make a generous enough offer to tempt Slovak decision-makers to abandon the dam. Most probably it did not even want to make that sacrifice for the future generations, since the results of the next elections would depend on its success or failure in coping with the grave short-term economic problems.

"Variant C" put the Hungarian government into an even more difficult situation. For a long time it committed the same mistake as Danube Circle: it did not take the Slovakian intention to go ahead with the construction of this variant seriously. However, with the new dam, supply of the old river bed with water fell completely into the hands of Slovakia, and Hungary is now at the mercy of Slovakia concerning the amount of water discharged into the Danube. The Hungarian dam-builder lobby immediately pointed to this fact as a result of Hungary's withdrawal from the treaty, since this would not have happened if the country stuck to the original contract. "Variant C" is also even more harmful to the environment than the original plan would have been.

As mentioned above, Hungarian government officials do not unanimously agree with the Parliament's decision and the government's official policy. Many of them share in the mentality of the dam-builders or even are linked to them. Frustrated government officials accuse Slovak nationalists just as much for the stalemate as they accuse the decision of the Hungarian Parliament, and especially the environmentalists, who put pressure on the decision-makers and made a compromise impossible. The Hungarian minister for the environment, S. K. Keresztes argued in an interview that the dam was not the only reason for the environmental and water shortage problems of Szigetkoz and Csallokoz (the areas affected by the dam system in Hungary and Slovakia), and environmentalists did not understand that only water engineering works could solve the problems there.   He accused the Danube Circle for preventing discussion of constructive solutions.[25]

## Austria

The Austrian government committed itself to back the Austrian companies which made a contract with Hungary in order to complete construction at Nagymaros with the participation of these firms.  The Austrian credits that were provided for financing the project enjoyed state guarantees on both sides.   At that time both the Hungarian and the Czechoslovakian government intended to go ahead with the construction.  In spite of this, the Austrian decision was quite controversial, since in both countries the regimes were non-democratic, and it was difficult not to notice public opposition to the project, at least in Hungary.

However, the contract provided jobs to Austrian construction firms and workers, cheap electricity, and huge profit for the powerful Austrian dam-builder companies which had been unable to start similar constructions at home, due to public protest.  The Austrian government received a lot of criticism because of its decision from Austrian environmentalists (and environmental groups of other countries as well), and politicians of the opposition in the Austrian Parliament.  Therefore the Austrian government was more cautious when Austrian financial participation was sought by the Slovakian government for the construction of "Variant C".  This time the position of the Hungarian government had also be taken into account. Austria did not want to damage its relations with Hungary, but it wanted to maintain equally good relations with Slovakia as well.   The Austrian government did not want to prevent its companies from making business in Slovakia, but did not want a conflict with its own environmentalists either. Therefore it did nothing:  it did not support Austrian companies which participated in the construction of Gabcikovo in any way, declared that the Austrian government had nothing to do with their business, at the same time making it clear that it had no legal powers to do anything against their activities.

Probably several government officials agreed silently with the Austrian businessman H. Androsch, who threw light upon certain Austrian considerations at the energy conference mentioned above:  Austria completely gave up nuclear power stations and limited the construction of hydroelectric plants as well, thus became a net importer of energy. Considerable amount of electric energy is imported from the nuclear power stations of the neighboring countries. The hydroelectric plant at Gabcikovo would contribute to a healthier ratio between nuclear and hydroelectric power in Slovakia. He blamed the Hungarians for forcing "Variant C" upon the Slovaks; this did not only increased the costs of construction, but also reduced the annual energy production capability of the Gabcikovo plant from 3800 to 2150 gigawatthours.    He also showed interest in the privatization of the Slovakian energy industry, since 49 percent of the shares can be bought up by the private sector, including foreign nationals.

## Legal Issues[26]

### The Legal Debate

Although both governments spoke of the necessity to treat the conflict as a professional debate which should be solved by "experts" (i.e., engineers, water management specialists and ecologists), attempt at solution in expert committees constantly failed, since the experts of the two sides were unable to agree about even basic facts.  After a while the debate continued among international lawyers of the two parties.  This section briefly describes the legal arguments of the two sides.

The debate is focused around two issues: 1. Is the unilateral termination of the treaty by Hungary acceptable from the point of view of international law?    2. Is the unilateral Slovakian solution ("Variant C") a legally acceptable response?

Since the 1977 treaty does not contain termination provisions, Slovak legal experts argue that Hungary's unilateral act to terminate the treaty constitutes a wrongful act.  They refer to the general rule of 'pacta sunt servanda', according to which both sides had to fulfill their obligation arising from the treaty in good faith.

However, international law permits the termination of treaties even if the treaty did not contain rules for termination, but only under certain circumstances strictly confined by international law.  During the negotiation process, and also in its declaration on the termination of the treaty, Hungary has presented five reasons why it considers its decision to discontinue the project not to constitute a violation of international law.  First, it stated that the 1977 treaty itself violated international law, or was derogated later by other rules of international law, such as the Stockholm Declaration, or the World Charter for Nature.  Second, it referred to the principle of fundamental change of circumstances (political changes in both countries; the dissolution of Council For Mutual Assistance; increasing importance of environmental considerations; new scientific evidence about the ecological impacts).  Third, it has invoked the principle of state of necessity (serious and imminent threat to vital ecological interest of Hungary) and fourth, the principle of moral impossibility (implementation of the treaty would be self-destructive for Hungary).  Fifth, it has argued that Czechoslovakia had committed a material breach of the 1977 treaty by not complying with those articles of the treaty which obliged both parties to ensure that the quality of the water in the Danube would not be impaired as a result of the construction and operation of the dam system.

A detailed description of the legal dispute would go beyond the purpose of this study. However, based on the opinion of independent legal experts, it is at least questionable to what an extent the Hungarian arguments hold water from the point of view of international law. These experts argue, that the treaty was not in conflict with binding international law. The Stockholm Declaration and the World Charter for Nature do not constitute binding legal principles that could derogate the 1977 treaty. The change of the political system or the change of political opinion, like growing environmental awareness, is generally not considered by international lawyers as fundamental change of circumstances. Concerning new scientific evidence, it was already available when Hungary and Czechoslovakia modified the agreement in February 1989, through which the Hungarian government implicitly confirmed once more its intention to participate in the project. Since Hungary deliberately decided to suspend the construction, and the implementation of the treaty did not endanger the very existence of the country, we cannot speak of moral impossibility either. Hungary could only successfully invoke the notion of state of necessity, because of the enormous negative impact of the project on the drinking water reservoir and on the ecology in general, and because it could be avoided without impairing essential interests of Czechoslovakia. Czechoslovakia's (Slovakia's) legitimate interests could be met by alternative solutions, and Hungary offered compensation which could cover the implementation of those alternative solutions. However, international courts have almost never had the opportunity to judge on this principle, and it is difficult to predict what decision they would take. Hungary could also insist that Czechoslovakia did not meet the environmental provisions of the treaty (most of the sewage treatment facilities that should have been built on the territory of Czechoslovakia, have not even been designed yet), but it is questionable whether this would be accepted by courts as justification for termination of the treaty.

Concerning the issue of "Variant C", Czechoslovakia argues that the unilateral solution became necessary because Hungary violated the treaty, and therefore it is a lawful reaction by Czechoslovakia, through which Czechoslovakia can mitigate damages caused by the wrongful act of Hungary. Czechoslovakia also claims that this solutions is only provisional, until Hungary continues the original project. Based on the opinions of the same independent international legal experts, Czechoslovakia's position is not justifiable from the point of view of international law. The 1977 treaty foresaw joint construction, operation and utilization of the project, but did not contain any provision for unilateral solutions. "Variant C" is a unilateral project, and Hungary has no longer the possibility to participate in controlling its operation or act at the locks of the head-water station in case of emergency. Therefore "Variant C" is in violation of the 1977 treaty, as well as other agreements, such as the 1976 Water Management Treaty, the Peace Treaty of Trianon, the Paris Peace Treaty and the 1956 Boundary Treaty (these treaties determined the frontier line between the two countries). "Variant C" also violates the principle of good neighborliness, because it causes significant harm to Hungary, and the principle of equitable utilization of the river. Even if we assume that Hungary violated the 1977 treaty by its decision to discontinue the project, it does not justify "Variant C" because it constitutes a violation of other rules of international law.

It is true that customary law recognizes unilateral reaction, such as retortion and reprisal. Retortion is an unfriendly act towards the injuring state, however, it must be in accordance with international law. Therefore, "Variant C" cannot be regarded as retortion. Reprisal may be an act violating a rule of international law, but its aim must be to obtain compensation or to force the other state to discontinue the violation of international law, and it must be proportional to the violation of international law by the other state. However, Hungary made it clear that it was not willing to continue the project, but was ready to pay compensation. "Variant C" is also an act which is out of proportion, because it causes damages to Hungary that might be irreversible.

## Attempts At Conflict Resolution

International law obliges states to find peaceful methods for dispute settlement. There are several ways of peaceful dispute settlement, such as negotiations or consultations between the parties; requesting a third party to participate in the discussions and serve as a mediator; referring the matter to a fact-finding committee; referring the issue to arbitration; referring the matter to the International Court of Justice. As we have seen, several of these methods have been tried.

Unfortunately, bilateral negotiations and consultations were not helpful in solving the conflict. The two parties could not even agree about basic facts. Therefore the establishment of a fact-finding committee became necessary, with the involvement of a third party (the Commission of the European Communities). Hungary would like to refer the case to the International Court of Justice (ICJ), but it only has jurisdiction over cases between states if they have agreed to refer the matter to the Court. The trilateral negotiations might result in such an agreement between Hungary and Czechoslovakia (from the end of 1992 the agreement must be made with Slovakia) if they can agree exactly on the issue to be brought to the Court. Based on the above analysis, it is understandable that Slovakia would like to refer the issue of unilateral termination of the 1977 treaty by Hungary to the ICJ, while Hungary is more interested in the ICJ's decision in the issue of "Variant C". However, Hungary is willing to submit the whole issue to the Court.

## Conclusions

The GNBS case is an exceptionally interesting environmental conflict to analyze because the three countries involved have been so different politically during the period examined. Austria represents a Western democracy, while Czechoslovakia and Hungary were communist countries at the beginning of the story, and both became pluralist democracies by the end. But the pace of democratization was very different in the two countries: in Hungary it was gradual, starting around 1985 and speeding up only after 1987, while in Czechoslovakia the change was sudden and radical, like an explosion, in December 1989.

The conflict had similar elements in all the three countries, regardless of political system. As it was demonstrated in the stakeholder analysis, the composition of groups supporting the project was similar in all the three countries, just as the composition of groups opposing it. Transnational cooperation among both the opponent and supporter forces of the three countries could also be observed. However, differences in the political systems of the three countries resulted in different political behaviors during the history of the conflict.

Although we can find economic short-sightedness and ecologically disastrous projects in all political systems, the non-market economies of the communist regimes put fewer limits to ruining the environment, because even economic rationality was overruled by ideological considerations. Further, important decisions were made without social control; information was monopolized by a small group of people, and society at large had no chance of taking part in the decision-making process. The possibilities of citizens for criticizing the decisions of the political leadership, organizing themselves in order to express and defend their interests were very limited or non-existent. Their efforts to express views different to the official standpoint were regarded as hostile and politically dangerous even in the case of non-political subjects. Actually, all such controversies became political, because they challenged the legitimacy of the system. Environmental protest played an important role in the democratic political change in Eastern Europe. In order to be able to defend effectively the citizens' rights to a healthy environment, environmentalists were forced to fight for freedom of speech, freedom of association, a free press, political pluralism and democracy.

Nevertheless, democracy is not a panacea for environmental problems. In parliamentary democracies, interests are represented through political parties, but for most of them -- except perhaps for the Greens -- other considerations are often more important than the environmental ones. If they support (or give in to) an environmental cause, they do so mainly for political considerations. The behavior of the Austrian government in the GNBS case is very illustrative of this: while it was forced to retreat in face of the citizens' protest in the case of ecologically risky projects on Austrian territory, it did not hesitate to finance and support similar projects in countries where the citizens' rights were suppressed.

If we examine which of the governments was the most sensitive to public pressure, we find that it was the Hungarian reform communist government of Miklos Nemeth. The reason for this was the unique political situation of prolonged transition from communism to market economy and democracy. The reform communists hoped to remain in power, or at least share power with the democratic forces after the elections. They also understood the grave mistakes committed by the communists regime and many of them believed in an improved, democratic Socialism. Therefore they gave way to public will more readily than the previous governments, or even their democratically elected heirs (for whom democratic elections gave such a legitimizing power that they could often disregard public opinion more easily). They also recognized how useful tool the GNBS case could become in their fight for power. Such a prolonged transition period did not take place in Czechoslovakia.

In Czechoslovakia the communists system was so oppressive that the opponents of the scheme could have no serious impact on decision-makers. The difference in pace between the political processes of the two socialist countries added an extra dimension to the conflict: the more open environmental and political protest in Hungary endangered not only the completion of the GNBS scheme, but also the regime itself. The case strained the already tense relations between the two countries -- resulting from the growing difference between their political systems -- even further. The activity of the Hungarian environmental movements encouraged corresponding movements in Slovakia. The Czechoslovakian party leaders were afraid of the power of example -- and they were right. The example of other countries became an important factor in the democratic revolution in Eastern Europe. Since real conflict between civil society and the central leadership could not evolve in Czechoslovakia due to repressive measures, the conflict appeared as an international one: it seemed as if menace to the regime came from outside, not from within. This obstacle in the way of solving the conflict was removed by the political changes.

However, although both Hungary and Czechoslovakia have acquired democratic political systems, this did not make the solution of the conflict much easier. Both countries are in deep economic crisis, therefore it is extremely difficult for them to give priority to environmental considerations over economic ones. To prevent this, a well-organized environmental movement and high ecological awareness of the population would be needed. In the latter respect there has been little difference between the two countries. If we look at the environmental movements of the two countries in general, we cannot observe a striking difference either. However, the conditions of the anti-GNBS movements were very different in the Hungary and Slovakia, especially since the mid-eighties.

In Hungary the movement was able to align the democratic and national opposition against the project, and gained the sympathy of the majority of the population. The project became a symbol of Communism is Hungary in the eyes of the people, and it was impossible for the new democratic government to defend it. However, the same did not happen in Slovakia. Opposition to the project was oppressed till the very last moment of the existence of the communist regime. The communist government pictured the conflict in its press campaign as one between two nations, the Hungarians and the Slovaks, and the Slovak anti-GNBS movement as one consisting of mainly minority Hungarians who were hostile to the Slovak nation and served Hungarian interests. Thus the environmentalists were put in a morally difficult situation, and their cause seemed more questionable to the Slovakian population than the one of the Hungarian movement in the eyes of the Hungarians. The new, democratic Slovakian government continued the policy of its communist predecessor in this respect.

The fact that this conflict, basically domestic in both countries, became one between them, can partially be explained by their different levels of commitment to the project (according to their expected gains and losses). Hungary has been less committed, because it has had the most to lose through the project, which rather served Slovak than Hungarian interests. Slovakia has also had a lot to loose (economically and environmentally), but it has expected the gains to outweigh the losses: the scheme has suited age-old Slovak national aspirations, therefore from a nationalistic point of view Slovakia could only gain from the project. In this respect it has been a zero-sum game between Hungary and Czechoslovakia.

The linkage with the national issue made the conflict very difficult to manage. Historically the Slovak-Hungarian relationship has been a conflicting one. After 40 years of forced "internationalism", the revival of national feelings in both countries is a natural phenomenon. Further, the Slovakian aspirations for independence revived the historical suspicion towards Hungary in Slovakia. Some nationalists circles want to increase this suspicion in the belief that a common enemy was necessary to unite the Slovakian nation. At the same time, the new Hungarian government strives to defend Hungarian national interests and those of the Hungarian minorities in the neighboring countries. While doing so, it often forgets about the sensitivity of the neighboring nations, and some members of the Hungarian government have made statements that were easy to misunderstand. However, the position of the Hungarian government is weaker in this respect, because it feels the Hungarian minority is kept hostage in Slovakia, and this fact is limiting its freedom of action. A lot of hesitation on the Hungarian side recently can probably explained by this fact.

However, neither of the two governments has done enough to move away from the zero sum game, providing the other side with safe-facing alternatives, although the Hungarians did offer solutions to the Slovaks such as building power stations operated with natural gas in Slovakia. However, the Slovak leaders have been suspicious of such solutions because they do not want that the energy supply of the new independent Slovakia would depend on other countries. It seems that it has equally been important for both governments "to look good" in the eyes of the West, especially the European Community. The game of the legal debate, and to a certain extent the bilateral and trilateral negotiations as well, have been played for the Western "audience": both countries have tried to prove that they were more "European" and democratic than the other, respected the Western values such as international law or protection of the environment more than the other, and were more favor of peaceful negotiations than the other.

The Hungarian government has committed several tactical mistakes during the negotiations. The hesitation of the Hungarian government and the Hungarian negotiators encouraged the Slovakian dam-builders to go on with the construction. This hesitation can partly be explained by inexperience, a will to avoid conflict with Slovakia, and a concern for the fate of minority Hungarians in Slovakia. However, it can also be explained by a lack of genuine will by some government officials to withdraw from the project. Those who still hope that the project would be completed according to the original plans, have a strong representation in the present political power as well. Speaking about inexperience, the Hungarian government and the environmental movement, especially the Danube Circle, committed the same tactical mistake of not taking "Variant C" seriously for a long time, in the belief that it was impossible for the Slovaks to construct it, and its aim was to blackmail the Hungarian government. Perhaps a generous compensation offer, at the same time providing the Slovaks with an opportunity for a face-saving retreat, would have solved the conflict then. In any way, it would have been easier for the Hungarians to make a deal with the liberal Czechoslovakian federal government than it is going to be with the nationalist Slovakian national government.

All the above factors influenced the course of the dispute, which has basically been one of conflicting paradigms, using the words of Boldizsar Nagy.[27] The Hungarian government's position represents a long term perspective, a preservation oriented approach, which cares for future generations, and adopts a precautionary principle regulating prudent behavior in circumstances of uncertainty. The Slovak government's position represents a short term perspective, not wanting to protect resources to be consumed in the remote future, and believing that man, with the help of technology, is capable of fixing whatever it destroyed. One paradigm has the goal of sustainable co-existence with nature, the other has industrial growth and modernization. These incompatible paradigms explain why the Hungarian and Slovakian negotiators were not able to agree even about basic facts.

A conflict of value systems is one that is most difficult to solve. Since bilateral negotiations led to a stalemate, they continued in a trilateral framework, and now the conflict is expected to be solved with the help of international law. The way of involvement of the Commission of the European Community shows that development philosophy still seems to prevail among experts and politicians. It is also questionable whether international law is prepared to protect the interests of long-term survival and sustainable use of resources. As we have seen in the section devoted to legal issues, the outcome of the case is most uncertain from the point of view of international law. Hungary seems to have more chances to win the case if the issue is "Variant C"; however it seems to have less chances concerning its unilateral withdrawal from the 1977 treaty. While treaty law has been in use and refined for a long time, international regulations protecting the environment have a much shorter history, and most of them are not even binding. Therefore it is highly questionable whether international law is equipped with the tools of protecting the environment and the survival of mankind. While there have been several cases in which the postmodern paradigm of sustainability won over the paradigm of industrial growth within a country, a similar decision by a court in an international dispute would be a new and important precedent for the future.

## Notes

1.  The World Rivers Review, a periodical of the International Rivers Network, devoted a special issue to large dams in 1991 (Volume 6,

Number 3, May/June 1991), on the occasion of the 17th ICOLD (International Commission on Large Dams) Congress in Vienna in June 1991. In this issue Leonard Sklar and Phil Williams list 12 problems dam builders cannot solve, such as reservoir sedimentation, coastal erosion, downstream channel erosion, increasing flood damages and risk, aging and decommissioning of dams, reservoir water quality and disease, soil waterlogging and salinization, safety problems, earthquakes, unpredictable flows, poor operation of reservoirs, and underestimating costs. Other articles describe protest against the Pak Mun Dam in Thailand, a national coalition in Brazil to stop dams, protest against construction of Icha Dam in India, a blockade to halt construction at the Czorsztyn Dam in Poland, and the emptying of the Nove Mlyny reservoir in Czechoslovakia, Southern Moravia, due to increasing criticism of environmentalists. Other issues of the journal describe several other examples from all over the world.

2.  Goldsmith, E. and Hildyard, N., 1984. *The Social and Environmental Effects of Large Dams*, Volume I. (Volume II: 1986). Wadebridge Ecological Centre, Cornwall.

3.  A good summary of the arguments, illustrated by examples, can be found in Williams, P. B., "The Debate Over Large Dams -- The Case Against", *Civil Engineering*, August 1991.

4.  This study is a revised, updated and largely rewritten version of a study written by the author in 1990: Galambos, J., "Political Aspects of an Environmental Conflict: The Case of the Gabcikovo-Nagymaros Dam System", in: Kakonen, J. (ed.), *Perspectives on Environmental Conflict and International Relations*, 1992, Tampere Peace Research Institute, published by Pinter Publishers Limited, London and New York.

    The author would like to express her thanks to Janos Vargha, who generously shared the rich documentation of ISTER (Institute for East European Environmental Research) about the case with her.

5.  This chapter was based on information gained from Kien, P. (pseudonym of Janos Vargha), "A nagy szlovak csatorna", *Beszelo, 9,* 1984, samizdat; Fisher, D., "Public intervention in pollution aspects of transboundary watercourses and international lakes. European experience.", manuscript, 1989; Fleischer, T., "Capafogsor a Dunan: a dunai vizlepcso esete", *Tarsadalomkutatas, 2/1992.*

6.  Vargha, J., "Egyre tavolabb a jotol", *Valosag, 11/1981.*

7.  Uj Szo, Bratislava, February 12, 1982.

8.  An expert committee was commissioned by the Nemeth government to prepare a background study for decision-making. The committee was lead by Peter Hardi, director of the Foreign Policy Institute in Budapest. The committee consisted of Hungarian, Slovak, Czech, Austrian, German, and American experts. They prepared a cost-benefit analysis of the project, according to which its complete cancellation was more favorable in the long run from an economic point of view. For the Hungarians, even in the short run, the costs of cancellation and continuation were estimated to be about the same. Their report suggested that a solution could also be found for the international legal aspects: compensation agreement or litigation between Austria and Hungary; a modification of the Czechoslovak-Hungarian agreement was recommended to be initiated, with reference to the points that stipulated the preservation of water quality and other environmental guarantees which have not yet been implemented. It was suggested that a bilateral agreement could be reached in terms of the net difference of gains and losses caused to both sides by the abandonment of the project. Hardi, P. et al., "The Hardi Report. Summary for the Council of Ministers of an expert review concerning the ecological, environmental, technical, economic, international and legal issues of the Bos-Nagymaros Barrage System", Budapest, 1989.

9.  Josef Vavrousek told about it in an interview with Cynthia Whitehead, journalist of *Environment Policy Europe*, on July 31, 1991.

10. Non-governmental organizations

11. Uj Szo, Bratislava, November 2, 1992.

12. In the summer of 1992, when it became obvious, that Variant C was being constructed, and that it had to be taken seriously, in spite of what Janos Vargha and Danube Circle had said, Andras Lanyi initiated the formation of the Danube Defense Action Committee. It consisted of the Danube Circle, Reflex, Ecoservice, Clean Air Action Group, and Slovak groups. The committee published a booklet *The Danube Blues: Questions and Answers about the Gabcikovo-Nagymaros Hydroelectric Station System*, and organized public protest against the construction of Variant C.

13. Bechtel was already approached by Hungary in 1988, to work as a consultant for the project. Bechtel was asked to develop an operation mode which would be an optimum solution between water management, energy production and ecology, giving priority to ecological considerations.

14. The press conference was held in the Presseclub Concordia, in Vienna, on October 22, 1992.

15. The historical chapter was based on the following materials:

Declaration of the Government of the Republic of Hungary on the Termination of the Treaty Concluded Between the People's Republic of Hungary and the Socialist Republic of Czechoslovakia on the Construction and Joint Operation of the Gabcikovo-Nagymaros Barrage System, Signed in Budapest on September 16, 1977, handed over, accompanying a note verbal, to the Embassy of the Czech and Slovak Federal Republic in Budapest on the May 19, 1992.

Protocol of the negotiations between the governmental representatives of the Hungarian Republic and the Czech and Slovak Federal Republic on issues related to the Gabcikovo-Nagymaros Barrage System on September 6, 1990 in Bratislava and on October 17-18, 1990 in Budapest.

Kien, P. (pseudonym of Janos Vargha), "A nagy szlovak csatorna", *Beszelo, 9*, samizdat, 1984.

Fisher, D., *Public intervention in pollution aspects of transboundary watercourses and international lakes. European experience*, manuscript, 1989.

Position of the Environmental, Economic, and Foreign Affairs Committees of the Hungarian Parliament on the hearing of Josef Vavrousek, Environmental Minister of the Czech and Slovak Federal Republic on September 11, 1991.

The resolution No. 104 of the Environmental Committee of the Czech and Slovak Federal Parliament on the problems of the Gabcikovo-Nagymaros Barrage System, adopted on September 18, 1991.

Resolution No. 239 of the Slovak Parliament adopted on January 31, 1992.

Nagy, B., *A Hungarian Chronology of the Bos (Gabcikovo) -- Nagymaros Dam System*, manuscript, 1991.

Hajosy, A., Hollos, L., *Damned Dams -- The Danube Story*, manuscript, Budapest, February 1991.

Mission Report of the Commission of the European Communities, Czech and Slovak Federal Republic, and Republic of Hungary, Fact Finding Mission On Variant C of the Gabcikovo-Nagymaros Project, Bratislava, October 31, 1992.

Leaflets of WWF Austria, SZOPK, Eurochain, the Danube Circle, Bela Liptak; Hungarian and English translations of Slovakian expert opinions and newspaper articles; numerous newspaper clippings from Hungarian and Slovakian newspapers (especially *Uj Szo*, which is published in Hungarian in Slovakia).

16. A detailed analysis of this problem can be found in Williams, P.B., "The Debate Over Large Dams -- The Case Against", *Civil Engineering*, August 1991.

17. Bela Borsos gave a detailed analysis of this mechanism in *Dams and Reforms in the East*, Budapest, 1989, manuscript.

18. See e.g. *Vodni hospodarstvi* 4/92, a journal published by the Federal Ministry for Environment in Prague.

19. More information about the Austrian dam-building industry can be found in Wiederstein, A. and Svarstad, H., "Dams: the Industry of Power", in *World Rivers Review*, 6, No. 3, May/June 1991.

20. For example, the Ataturk dam in Turkey, which has such undeniable potential as a strategic weapon, that even the World Bank refused to get involved in the project. Designed with the capacity to turn off the entire water flow of both the Tigris and Euphrates rivers, the project gives Turkey the ability to control the main supplies of fresh water to both Syria and Iraq. Quoted from Wiederstein, ibid.

21. See Hardi P. et al, The Hardi Report, ibid.

22. Szabo, M., "Vannak-e alternativ tarsadalmi mozgalmak Magyarorszagon?" <Are there alternative social movements in Hungary?>, in: L. Solyom and M. Szabo (eds), *A zold hullam* <*The Green Wave*>, Eotvos Lorand Tudomanyegyetem Allam- es Jogtudomanyi Kar, Budapest, 1988.

23. See a more detailed analysis in Fleischer, T., "Capafogsor a Dunan: a dunai vizlepcso esete" <Jaws On the Danube: The Case of the Middle Danube Hydroelectric Dam>, *Tarsadalomkutatas* 2/1992, Budapest.

24. The following historic explanation was first published in samizdat, by Janos Vargha, under the pseudonym Peter Kien. Beszelo, 1984/9.

25. *Kisalfold*, May 11, 1992.

26. This section was based on the following literature: "Construction and Operation of Variant C of the Gabcikovo-Nagymaros Project under International Law", Legal Study for the World Wildlife Fund for Nature by Rechtsanwalt Dr. Georg M. Berrisch, LL.M., Brussels of Schon Nolte Finkelnburg & Clemm, October 1992;

Nagy, B., "Five Theses on the Legal Possibilities of termination of the Intergovernmental Treaty of the Bos(Gabcikovo)-Nagymaros Barrage System, Concluded in 1977", April 1992, ISTER, Budapest;

Nagy, B., "The Danube Dispute: Conflicting Paradigms", ISTER, November 1992, submitted to the New Hungarian Quarterly for publication;

Hunter, D., "The International Legal Aspects of Unilaterally Constructing the Gabcikovo Dam", Center for International Environmental Law, Washington, May 21, 1992 (draft);

Balas, V., "The Case of Gabcikovo-Nagymaros: International Legal Aspects", Institute of State and Law, Prague, September 1992;

"International law analysis of Option "C" -- Completion of the Gabcikovo Water Project on Czecho-Slovak Territory without Agreement with Hungary", Prague, October 29, 1990 (in English, no author's name on the manuscript).

27. Nagy, B., "The Danube Dispute: Conflicting Paradigms", ibid.

Chapter 10

# ENVIRONMENTAL CONFLICT AND POLITICAL CHANGE: PUBLIC PERCEPTION ON LOW-LEVEL RADIOACTIVE WASTE MANAGEMENT IN HUNGARY*

Judit Juhasz, Anna Vari, and Janos Tolgyesi

In this paper, in connection with a case study, we would like to illustrate how environmental problems emerged in Hungary, what the response of the general public was to such problems, how environmental movements became organized, and how all this fit into the major political changes characterizing the period between 1985-1990.

## Background

For about forty years, Eastern European societies had ideologically denied even the theoretical possibility of environmental problems. They assumed socialism kept economic development and the environment in harmony. With hyper-optimism many believed that rivers could be turned around and that human beings absolutely ruled nature. Human dominance over nature was regarded as the strength of socialism. According to the socialist slogans, nuclear energy was in service of peaceful production, and nuclear risk could only be mentioned in connection with the weapons of the imperialists.

From the late 1970s, environmental problems started to be oppressive in Eastern Europe. In Hungary, for example, it became obvious that in several regions, groundwater and air pollution was extensive and extremely high. Clearly, the environment had become a major source of danger, and many indices of the health conditions of the population had undoubtedly developed in a lasting, adverse manner. In response to these dangers, environmental groups emerged, most of them with sincere sensitivity towards local or regional environmental issues (Vari & Farago 1991, Szirmai, this volume).

In 1985, parliamentary and municipal elections took place in Hungary, and were based on somewhat more liberal rules than formerly. Many candidates of the opposition put environmental issues on their political agendas. Representatives of the opposition were not particularly successful in the parliamentary elections, but on the local level significant changes took place. In many municipalities, unpopular former officers failed in their bids for re-election and people from the opposition were elected. This experience was encouraging for the public, and signaled the possibility of intervention and changes.

*Public attitudes outlined in this paper were partially derived from a public opinion survey conducted by the Hungarian Institute of Public Opinion Research. We gratefully acknowledge the assistance of Klara Farago, Katalin Solymosi, and Terez Wekler in conducting the survey.

A. Vari and P. Tamas (eds.), Environment and Democratic Transition, 227–248.

The presented case is associated with the attempts to site a low- and intermediate level radioactive waste repository between 1976 and 1990. The controversy about the siting illustrates the declining phase of the political regime.   First, the history of the siting is presented, followed by the description of public attitudes concerning the planned facility.  The study is based on (i) the analysis of relevant documents, including regulations, minutes of public meetings and expert debates, (ii) semi-structured interviews conducted with key actors of the siting process between June and October 1988, and (iii) a public opinion survey carried out in May 1989.

## The History of the Siting of a Radioactive Waste Repository

### Chronology of the Main Events

The problem of the disposal of low- and intermediate level radioactive waste first emerged in 1976 during the construction in Paks (Tolna County) of the first nuclear power plant.  It became clear that the waste of the power station could not be exported to the Soviet Union and the power station had to dispose of it within the country.   The first alternative was enlarging an existing belowground repository established by the Ministry of Health for low-level radioactive medical waste disposal.  This option, however, was not endorsed by the government and the nuclear power station decided to establish a new belowground disposal facility.

A legal and regulatory framework for establishing a new disposal facility had never been elaborated.  It was not clear which bodies were responsible for regulation and licensing, and according to what rules.  Permits had to be obtained from county councils, and a series of other permits were needed from other authorities.  Issuing the operation license was the task of the Ministry of Health, but its decision could be highly influenced by other agencies.

The contractors of the power plant identified eighteen alternative sites, and from this set, the site near Magyaregregy (Baranya County) was chosen in 1983.  For further testing, getting a permit from the county council was necessary.  This was supposed to be a routine administrative procedure.  The county council, however, refused to issue the permit, based on the negative opinion of a group of local experts about geological conditions.

In 1984, experts of the power station suggested another village, Ofalu, also located in Baranya County as a site.  This time, the leaders of the power plant tried to win over some central governmental agencies to the siting.  They were successful:  due to the pressure of the central government, the county council issued a conditional permit for the testing.  Local experts again opposed the site.  They argued that the soil was not impermeable, there were some wells nearby, and the region was seismologically active.  They claimed that under certain circumstances, radioactive substances could be released into the environment. The leadership of the power plant did not consider this opposition to be a serious threat to their plans. They assumed they could prove the suitability of the site with new geological tests.

Late in 1987, preparations for testing started, and shortly thereafter, the news about the siting leaked out. Local residents began to oppose the siting. In response, the power plant organized public meetings in February and March 1988, where they briefed the local population about the disposal technology and the site selection process. The information, however, came too late. Far from assuring the residents, the briefing made them even more suspicious. They felt the briefing wasn't of appropriate tone and that portions were obscure or contradictory.

The residents started to organize their resistance. In March 1988, a citizens' committee was formed which established a panel of independent experts. The panel was charged with the task of evaluating of the results of the previous geological tests. For the first time, the leaders of the power plant realized that they would have to take the residents more seriously. They offered compensation and some degree of control over the operation of the facility, but all this came too late. The residents asserted that their acceptance would depend on the opinion of the independent experts.

Having investigated the geological data, the experts found the site technically inappropriate. In April 1988, a public debate was organized between the experts of the power plant and the experts appointed by the residents. Because the opposing groups could not reach a consensus, the county council did not issue the land use permit. The power plant's appeal to central governmental authorities was also rejected, and in the summer of 1988, the siting was suspended.

In the fall of 1988, the state administration tried to get out of this stalemate by requesting that a committee of the Hungarian Academy of Sciences evaluate the arguments of the two opposing expert groups. Even in this situation, the power plant leaders did not investigate the possibility of alternative sites or disposal technologies. They formed the opinion that after the events to date, the disposal facility would not be accepted in any other place, and implementing a safer technology, like underground mined disposal, would be very expensive. Further, they seemed to regard giving up their original ideas as incurring an intolerable loss of prestige. They tried to convince the residents again and repeated their offer of compensation.

The year 1989, however, was already the year of basic political changes. The movement against the siting came to be interpreted as a struggle of the local population against the central power. In the spring of 1989, the representatives of the affected villages decided to refuse the offer of the power plant and protested with a large demonstration against the siting. Finally, in June 1989, the committee of the Academy of Sciences formulated its position in a rather ambiguous manner. They stated: "The site is not inappropriate." Based on this viewpoint, the Ministry of Health did not issue the license. In October 1989, the power plant appealed to the Council of Ministers for a reconsideration of the case. The government, however, focusing on the approaching elections, rejected reconsideration of the previous decisions.

## The Social Context

The village short-listed as the site of the disposal facility was Ofalu, a village of approximately 500 inhabitants, in the South Western part of the country, 80 km from Paks, but in another county. Administratively, Ofalu belongs to Mecseknadasd, a village of about 1,700 inhabitants. This means that most public institutions serving Ofalu (and two other nearby villages) are located in Mecseknadasd. The ratio of intellectuals and other educated people in this village is relatively high. The standard of living is also higher than the average in similar settlements.

The population of Ofalu, Mecseknadasd, and some other nearby settlements is mostly ethnic Germans. Interviews suggest that because of the deportation of many German residents after World War II, and the pressure which followed, there had always been a strong fear by the ethnic Germans of those in power. Many of the villagers were not enthusiastic about the regime and its representatives. Before 1985, most were not appreciative of the council chairman of the village, who was not a local man and who, according to the public, did very few for the village during his 20 years in office.

By the 1980s, a generation that had grown up in the liberal Kadar era became active. Many of them were educated outside the village or worked in nearby towns. Young people, including those who studied somewhere else or worked outside the village, regularly met in the nationality dance group of the village. This is where they started to organize themselves to ensure that their own candidate would win the 1985 local elections. They were successful; the new council chairman was a member of the dance group, a young intellectual, who together with his wife had been born in the village. He started to work with great intensity and was able to implement development which was unparalleled in the past. Interviews suggest that people appreciated his knowledge, his work, and his honesty.

## Resistance to the Siting

In the fall of 1987, news about the siting leaked out. This happened a year after the Chernobyl accident and nuclear danger was very much alive in public opinion. A small group of intellectuals started to organize opposition to the siting. The main organizer was the wife of the council chairman, a sociologist who worked in the nuclear power plant. On her invitation, the representatives briefed the residents of Ofalu and Mecseknadasd in the spring of 1988. The briefings, however, became counterproductive. Interviews suggest that the exaggerated self-confidence and the condescending tone irritated people. The argument that the technique to be applied in the facility is safe did not convince them. People did not trust the products of the domestic industry, nor the experts involved in the siting and operation of the facility. On the other hand, people recognized that a decision had been taken without informing them. This created further doubts (for if it is free of danger, why did they not speak earlier about it) and the decision-making process itself provoked resistance.

The organizers of the resistance primarily endeavored to utilize formal contacts to prevent the final decision on the siting of the disposal facility. They tried to obtain information from state and county officers, while mobilizing the population and the leaders of other affected settlements. For the activists, the stake was not only the protection of the environment, but also the indication that expressing and implementing local interests was possible.

According to interviews, none of the promoters of the siting seriously counted on the fact that the initiative of the Mecseknadasd people could interfere with the procedure. The designer continued the professional preparation, the plans of the project were completed, the bureaucratic procedures associated with the permit went on their way, and several approvals by the authorities had already arrived. In March 1988, six months after the first briefings, the movement found that it was not taken seriously. Therefore, a few persons in the leadership of the resistance decided to mobilize the media.

At the time, such topics started to interest the media. Many journalists associated with the local, county and national press, sympathized with ecological and environmental efforts or strived to discuss local issues and undertook to write about environmental conflicts. Later, interviews and reports were published in the national and local radio and on the television. The media raised the siting into a national issue and permanently kept it on the agenda.

The media featured the case of the little ethnic minority village, Ofalu, oppressed and deprived, as the youngest prince in a fairy tale, who came from the unknown and defeated the fearsome foe, the invisible villain or the giant, which at first reading is the danger of the nuclear waste, but ultimately is the central government.

After the appearance of the first articles in the national and county dailies, even the earlier quiet population became encouraged to take a more open stand. Whereas earlier, open opposition took place only in Mecseknadasd, now the residents of the other villages started to join in. The winning over of the media gave a significant impetus to the case.

## The Institutionalization of the Resistance

The leaders of the opposition to the facility attempted to find an appropriate structural and legal form in order to gain a substantial position for negotiation. They initiated the formation of an association, but the permit for this was delayed, because of administrative and political obstacles. Thus, at a village meeting, the setting up of a citizen committee was decided upon in March 1988. Each affected village nominated one member to the committee. The committee established a panel of independent experts to review existing research materials and documents concerning the suitability of the site. The negative viewpoint of the panel came out in April 1988. It provided the resistance with professional arguments: they published their fears in the form of a petition, which was signed by most in the concerned settlements and thousands more in the county seat.

The county administration did not know how to react to the vehement opposition. The county leadership was not united. The protest of the residents had supporters as well as opponents in the executive apparatus and the party leadership. As no similar case had occured in the past, the decision makers felt confused, and did not want to take a stand on either side. The shortcomings of the regulation came to the surface. County officers placed their hopes on the dispute of the experts, but this did not lead to any agreement, so in the summer of 1988, the county government suspended the siting.

Even on the level of the central government, nobody wanted to undertake the responsibility for a decision. A further procrastrination started. Finally, in the fall of 1988, the Hungarian Academy of Sciences was asked to form an opinion. With this, the case was shelved. Both the power plant leaders and the local population counted on a favorable opinion from the Academy. The plant leaders also hoped that during this interval the waves would subside. Events, however, turned out differently.

## The Political End Game

In the spring of 1989, many new political organizations emerged. The movement against the siting obtained symbolic value, and the media continued to autopsy the case. The press support and public opinion did not let the matter die down. The standpoints of the main antagonists hardened.

The plant leaders did not see any other way out than by offering compensation to the public. They made their offer directly prior to the publication of the Academy statement, when news leaked out that within days a decision favorable for the village would be made. Not only was the timing of the action bad, but what they offered and why was not clear.

The population proved to be united. Their answer, unanimously decided at a village meeting, which was published as an open letter in the dailies, refused the offer as a "clear intention to bribe." The action had a boomerang effect because the press again aired the case. The shaky leadership of the central government, which wanted to prove its democratic reform nature, was not in a position to force the siting on the local people by administrative means.

The citizens' committee did not want further spectacular actions. They believed that the case was moving smoothly ahead. But the local organization of the Hungarian Democratic Forum -- one of the major opposition parties -- wanted to prove its alignment. It organized a national protest demonstration at the site, again creating great press and television publicity for the organizers. The population enthusiastically participated at the demonstration.

The Academy also felt its delicate situation. Its standpoint was summarized in a rather ambiguous manner: "The site is not inappropriate." They did not oppose the site for its technical aspects but added that "the scientific dispute is only secondary, compared to the basic frustration of the population." As the earlier negative decisions were supported by professional reasoning, the power plant leaders hoped that the opinion of the Academy would help them. This, however, was not the case. The license was not issued.

After this, there was only one opportunity left for the power plant. In the fall of 1989, it appealed to the Council of Ministers. Amidst the accelerated political changes, the decision was again delayed. Attention was drawn away. The media focused on the preparations for the election and the party battles.

In the villages of the Mecseknadasd region, there was hopeful expectation with regard to the election. No local party organizations were set up, but sympathizing circles appeared around the different opposition parties. The leaders were usually the same persons who were active in the siting case. The earlier acquired popularity of the council chairman became a commodity much in demand, and several parties approached him, offering him a candidacy. Finally, three candidates from the concerned area were chosen to stand for election, including the council chairman.

The final refusal concerning the siting of the waste disposal facility arrived at the climax of the election campaign, providing considerable advantage for the council chairman in the canvassing. His election slogan: "He who could win in Ofalu, can also represent our interests" -- was successful. He was elected to Parliament with a landslide victory and he became the head of the parliamentary committee for local government.

## Public Opinion Survey Concerning the Planned Radioactive Waste Disposal Facility

### The Means, Methods and Samples of the Survey

The Hungarian Institute of Public Opinion Research conducted a survey on the siting of the low-level radioactive waste disposal facility in May 1989. A standardized questionnaire was used to elicit attitude associated with the following:

>   Importance and causes of various environmental problems in Hungary,
>   Energy production, various sources of energy,
>   The siting of the radioactive waste disposal facility,
>   The experts involved in the siting,
>   The decisions and the participants in the decisions,
>   Compensation and incentives.

Based upon prior research, the study hypotheses included that the attitudes of the Hungarian public towards both nuclear energy production and radioactive waste disposal tend to be negative, primarily because the public perceives nuclear technologies as unsafe and does not trust the institutions regulating, developing and operating nuclear facilities (Valko 1989, Vari, Kemp, & Mumpower 1991, Vari & Farago 1991, Juhasz & Tolgyesi, in press). A number of Western studies, however, suggest that certain measures including compensation and incentives, public participation in the siting decisions, and public monitoring and control over the construction and operation of the facility may make attitudes towards a nuclear waste repository more positive (Johnson 1987, van der Pligt, in press).

According to several research studies (Covello 1983, Freudenburg & Rosa 1984, van der Pligt in press), the public's attitude to radioactive waste disposal facilities depends, to a great extent, on how directly they are to be affected by them. Communities close to the planned site are usually more concerned than communities of a greater distance. On the other hand, local acceptance is usually better in communities which host or are closely located to already existing nuclear facilities, where people are more familiar with the technology and enjoy direct economic benefits of these facilities (Earle & Lindell 1982, Morell & Magorian 1982, van der Pligt, in press).

To ascertain how far public attitudes were dependent on how directly the individuals would be affected by the planned low-level radioactive waste disposal facility, samples were chosen from three different regions: 300 from six villages in the Ofalu region (Baranya County), 200 from three villages in the area around the Paks Nuclear Power Station (Tolna County), and a further 200 from three villages in Vas County, a region which will be referred to as the neutral region, as it is far removed from any existing or planned nuclear facilities.

### Attitudes Towards Environmental Problems

Those interviewed were first requested to enumerate those problems which are of most concern in Hungary today. The responses showed that in 1989, environmental issues were seen as everyday problems: nearly a quarter of the respondents mentioned them.

The answers were clearly influenced by how respondents would be affected by the disposal facility: 39 percent of the respondents from the Ofalu region defined environmental problems as among the most pressing, compared with 13 percent of those living around the Paks Nuclear Power Station, and 21 percent of those living in the so-called neutral region.

With regard to the seriousness of environmental problems the responses followed a similar pattern. Eight problem areas[1] were outlined and respondents were asked to what extent (very much, quite a lot, not much or not at all) these were serious problems in Hungary today. Most of the respondents from the Ofalu region thought that environmental pollution was the second most serious problem (after inflation), whereas in the other regions, they were only ranked in fourth or fifth place.

In general, it was the young people, the educated, and the men who thought that environmental issues were important, but the importance of the social and demographic factors was moderate. The degree of concern about the environment was primarily determined by where the respondents were living.

In the Ofalu region, two other factors may account for a higher degree of environmental consciousness. First, most people living here are of German origin who may have been more aware of environmental issues through personal contacts with Germany than the Hungarian population. Second, villages in this region have been in a better economic situation, with relatively higher living standards and better living conditions than those in the other two regions.

Previous research suggests that people with a higher standard of living and better housing take environmental issues more seriously (Dunlap & Baxter

1988, Harsfalvi et al. 1989, Inglehart 1990). Indeed, the community most active in the fight against the disposal facility was Magyaregregy, with the highest standard of living. On the other hand, Bataapati, the only village in the Ofalu region which was willing to accept the facility, had a much poorer infrastructure and general standard of living that any of the other villages in the area.

## Assessment of Various Environmental Problems

Residents in the three regions were asked about the significance of various environmental problems. They were asked to indicate how serious they considered environmental problems to be in Hungary. The answers were based on a scale of 1-4, in which 1= not at all serious, and 4= very serious.

Respondents in the three regions significantly differed in their assessment of various types of environmental hazards (Table 10-1). For those living in the Ofalu region, accumulation of radioactive waste was the most serious of eight environmental dangers listed, with the accumulation of toxic industrial waste coming next. Not far behind came water pollution, one of the perceived impacts of the disposal facility. These three problems were seen as the most serious in the three regions but were weighted in different ways. In the neutral area these three problems were seen as more or less equal whereas in the Paks region, the pollution of lakes and rivers was seen as most threatening.

Table 10-1. The Evaluation of Certain Environmental Problems (the Three Regions and the National Average)

| | Ofalu | Paks | Neutr. | N | F | Sig. | National (1988)[2] |
|---|---|---|---|---|---|---|---|
| Accumulation of radioactive waste | 3.73 | 3.69 | 3.48 | 689 | 5.27 | 0.05 | 3.71 |
| Accumulation of toxic industrial waste | 3.56 | 3.59 | 3.47 | 696 | 1.02 | 0.03 | --- |
| Lake and river pollution | 3.55 | 3.77 | 3.48 | 696 | 8.08 | 0.01 | 3.59 |
| Air pollution | 3.40 | 3.57 | 3.41 | 697 | 2.89 | 0.05 | 3.63 |
| Accumulation of household waste | 2.97 | 3.18 | 3.17 | 697 | 4.54 | 0.01 | 3.27 |
| Noise pollution | 2.86 | 2.94 | 3.19 | 697 | 7.54 | 0.01 | --- |

Differences in evaluation of these problems were small between the various social and demographic groups. The results of further statistical analysis revealed that the factors of toxic and radioactive waste assessment were similar to each other and different from those of the other problems. The evaluation of radioactive and toxic waste was influenced primarily by the age of the respondents (young people were more concerned), but the response to the other problems was more determined by the general attitude to the environment -- primarily to what extent those questioned were interested in environmental issues. The number of children of the inhabitants was also a determining factor of the attitudes towards toxic and radioactive waste, which reflects a fear of what might happen in the future. During the debate in Ofalu, the phrase, "We cannot decide for our grandchildren," was repeatedly iterated as a reason for opposing the facility and refusing compensation.

*Reasons for Environmental Problems*

The fear expressed by the population and their attitudes may lead to the assumption that their opposition was not merely -- or even primarily -- prompted by anxiety about the waste disposal facility, but by a more general fear based on previous experiences. Indeed, acquaintance with the technically advanced design and safety procedures had very little effect on their attitudes. It was clear from the reactions to the arguments of the investors at the village meetings and from our interviews that even if the residents believed that the disposal technology was entirely safe they could not be persuaded that such a plan could actually operate safely in Hungary. This attitude has not as much to do with the facility itself as with general experiences and opinions. To get a fuller picture of these attitudes, we asked what factors are perceived to give rise to environmental problems. We listed four factors as follows:

1. Lack of money;
2. Behavioral factors (negligence, the failure to follow instructions);
3. Politics, questions of central control (political considerations, production motives, the lack of appropriate regulations); and
4. Local community factors (the community's inability to protest and organize, and lack of competence).

Most of the respondents listed lack of funds and behavioral factors as the main reasons for the problems. The settlements differed over two things. Lack of money was seen as the main reason for the problems by 40 percent of the respondents from the Paks and neutral regions, whereas in the Ofalu region only 28 percent of the respondents thought so. The other important difference was that nearly one-tenth of the respondents from the Ofalu region said that the local community factor was the most important whereas the others scarcely mentioned it at all. The behavioral factors -- which cannot be eliminated by money and technology -- were seen as the main reason for environmental problems by a third of the respondents and in this respect there was no difference between the three groups of settlements.

## Opinions about the Environmental Activities of Certain Institutions and Organizations

Interviewees were asked about the role of various institutions and organizations in environmental protection. They were asked the question: How much do you think the following do to protect the environment? The respondents were given four possible answers: 4= everything in their power, 3= quite a lot, 2= fairly little, 1= absolutely nothing.

The data show that the population has a fairly negative view of the environmental protection activities of the various institutions and organizations (Table 10-2). The majority believed that the central government, the Ministry of Environmental Protection, and the local councils do very little or almost nothing to protect the environment. The respondents' opinion of the economic bodies was even worse, but the majority also viewed unfavorably the local population's environmental activities. Of the list we composed (government, Ministry of Environmental Protection, local councils, general public, mass media, factories, businesses and collective farms) the mass media alone was perceived by the majority of the respondents as doing something for the environment. Three-quarters of the respondents said that the mass media did quite a lot or even everything in its power to protect the environment.

Table 10-2.  Evaluation of Environmental Activity

|  | Ofalu | Paks | Neutr. | N | F | Sig. |
|---|---|---|---|---|---|---|
| Radio, TV, Papers | 2.69 | 3.05 | 2.77 | 696 | 9.88 | 0.001 |
| General public | 2.51 | 2.25 | 2.15 | 697 | 10.06 | 0.001 |
| Local councils | 2.36 | 2.13 | 2.04 | 696 | 6.96 | 0.001 |
| Government | 1.85 | 2.21 | 2.03 | 696 | 10.71 | 0.001 |
| Min. of Environment | 1.83 | 2.22 | 2.13 | 690 | 9.94 | 0.001 |
| Factories, businesses, collective farms | 1.62 | 1.78 | 1.84 | 696 | 3.8 | 0.02 |
| Own activity | 2.83 | 2.78 | 3.00 | 698 | 3.75 | 0.02 |

Only one-third of those questioned in the Ofalu region knew about groups and organizations that we had not listed and which do a lot for environmental protection, although in the affected settlements strong local groups had been formed to protest against the disposal of radioactive waste. Elsewhere, less than one-fifth had any knowledge of such groups, although during this period several well-known environmental groups existed in Hungary.

*Evaluation of the Energy Situation, Attitudes to Nuclear Energy and Other Sources of Energy*

Energy supply touches on nearly all areas of life and is one of the most basic necessities. Problems that arise have a strong influence on the judgment of related areas, such as opinions about the building of power stations and the need to operate them. They, in turn, influence views on waste disposal. If the energy supply is threatened or the price of energy increases, questions of pollution and environmental hazards may become less important in the public's opinion.

In the last few decades Hungary has experienced a constant supply of energy which was relatively cheap, thus for a long time the population has not been concerned with energy supply problems and their consequences. During the period of the survey, mid-1989, the possibility of energy problems was often discussed in various forums, including the mass media. This resulted in an increased awareness of potential energy supply problems (Valko 1989).

In the sample, 60 percent of those living in the Ofalu area and more than two-thirds of those in the other areas said that they expected there would be energy supply problems in the near future in Hungary. Respondents were asked to evaluate the various solutions to the energy problem on a scale of 1 to 5, where 1 = very bad and 5 = very good. Opinions about practical ways to resolve the energy problems are shown in Table 10-3.

Table 10-3. Evaluation of Practical Ways to Solve the Energy Problems

|  | Ofalu | Paks | Neutr. | N | F | Sig. |
|---|---|---|---|---|---|---|
| Thermal Power Plants | 3.38 | 2.84 | 2.82 | 694 | 4.94 | 0.01 |
| Hydroelectric Plants | 3.21 | 3.15 | 3.29 | 693 | 1.61 | 0.1 |
| Nuclear Power Stations | 1.33 | 1.82 | 1.66 | 696 | 8.28 | 0.001 |
| Imported Energy | 2.41 | 1.96 | 2.39 | 695 | 4.66 | 0.01 |
| Solar and Wind Energy | 4.40 | 4.21 | 4.44 | 693 | 4.39 | 0.01 |
| Conservation in Industry | 4.37 | 3.95 | 4.34 | 696 | 5.91 | 0.01 |
| Conservation in Households | 4.13 | 4.17 | 4.26 | 691 | 1.11 | --- |

In all three regions the use of non-traditional energy sources, such as wind and solar energy and conservation were considered the best solutions to the energy problem. At the same time, it was generally believed that the building of nuclear power plants would be the worst possible solution. Of the respondents, those living in the Paks area were least opposed to constructing nuclear power plants whereas those from the Ofalu area were most opposed.

With regard to traditional energy sources, in the Paks and neutral regions, the construction of a hydroelectric power station was said to be the most appropriate way to solve the energy problem whereas many in the Ofalu region said that the construction of a thermal power plant would be the best solution. In the Ofalu region, the preference for a thermal power station may be related to the fact that the local thermal power station is a major source of local employment.

Demographic characteristics barely affected the evaluation of the various energy sources. The respondents' sensitivity to environmental problems was similarly insignificant in these responses.

### Knowledge about the Planned Facility

At the time of the survey the plan to create a disposal facility for the radioactive waste from the Paks Nuclear Power Station was well known. In the Ofalu region all respondents were familiar with these plans; in the Paks region 84 percent knew of them and even in the neutral region, which is relatively far away, more than 70 percent had heard about the facility. In the Paks region, 90 percent of those who knew about the facility could also name the planned site, whereas in the neutral region two-thirds of the respondents could do this.

The mass media played an important role in distributing information associated with the siting: 92 percent of those living in the neutral region and 75 percent of those in the Paks region said they had received their information from the mass media. Residents of the Ofalu region got most of their information from personal contacts: 50 percent first heard about the plans from relatives and friends, and 13 percent from workers employed on the site. One-fifth of the Ofalu region respondents said they had heard about it through local forums, whereas one-tenth said that they heard about the plans from the mass media.

### Attitudes to the Facility

The majority of those asked were against the facility being sited in the Ofalu area. The answers, however, varied according to the degree to which the respondents were affected. In the Ofalu region 94 percent did not agree with the selection of the site and a further five percent said their consent would depend on certain conditions. Even in the communities of the Paks region only 15 percent of the respondents agreed with the selection of the site, whereas 56 percent were against it and a further 29 percent were unsure. In the villages of the neutral region 62 percent of the respondents opposed the Ofalu site and 13 percent would accept it given certain conditions.

The acceptance of the site depended only to a small extent on demographic features. In the Paks area higher education played a significant role; more than 70 percent of those who had not completed 8th grade and only one-half of those who had completed secondary school disagreed with the site. To evaluate this, we must note that this latter group is the best informed about nuclear energy and often its responses were supported by rational considerations of the issue (e.g., they frequently stated that the waste must be disposed of somewhere and many of them referred to scientific and technical considerations, too).

Attitudes toward the siting and evaluations of the political situation were somewhat interrelated. In the entire sample, two-thirds of those who did not agree with the governments' policy in general opposed the siting, whereas only one-half of those who more or less supported the government were against the siting ($p > 0.005$). In the Paks region only 40 percent of those who supported the government disagreed with the selected site compared to 67 percent of those opposed to the government ($p > 0.01$). Political factors may have played an important role despite the fact that only a few of those against the siting explained their attitudes by political or social reasons.

With regard to the the long-term effects of the facility, no respondent in the Ofalu region regarded it as completely safe, whereas half of the respondents said that it was absolutely unsafe, and a further 30 percent were only a little more positive saying they thought it was not very safe. The responses in the other two regions were less negative. In the Paks region and in the neutral region 20 percent and 26 percent, respectively, said they thought the facility was absolutely unsafe. Besides, in the Paks region and in the neutral region 21 percent and 12 percent, respectively, said that the repository was fairly safe.

We listed four problems concerning the possible risks of storing radioactive waste and asked about their likelihood. We used the following scale: 4 = very likely, 3 = fairly likely, 2 = not very likely, and 1 = very unlikely. The results are displayed in Table 10-4.

Respondents from the Ofalu region estimated all of the enumerated risks higher than the other groups, whereas those from the Paks region estimated most risks -- except for that of a transportation accident -- lowest. It is worth noting that a transportation accident would indeed affect both the Ofalu and the Paks regions in a similar way.

Table 10-4. Likelihood of Certain Risks

| | Ofalu | Paks | Neutr. | N | F | Sig. |
|---|---|---|---|---|---|---|
| The radioactive waste escapes and pollutes the rivers | 3.59 | 3.12 | 3.15 | 698 | 17.0 | 0.001 |
| After an earthquake the the waste is exposed | 3.55 | 3.27 | 3.29 | 697 | 7.5 | 0.001 |
| Accidents happen during transportation | 3.03 | 2.94 | 2.78 | 698 | 4.97 | 0.01 |
| Our successors will accidentally uncover the site | 2.70 | 2.46 | 2.63 | 699 | 0.8 | --- |

These results again show a greater trust in technology and skilled personnel by those living in the nuclear power plant area. This is also reinforced by their greater trust in the safety of radioactive waste disposal in general: 40 percent thought it was possible to make it safe, whereas the number of those who felt the same way in the Ofalu region was 20 percent, and 26 percent thought so in the neutral region.

Demographic and social factors played a lesser role in the estimation of the likelihood of safe operation. The young people, the men and those who were more highly educated were more optimistic that a safe technology could be found.

Independent of evaluations of safety, hardly anyone wanted the waste to be disposed of in their area (5 km from their homes ), despite the promise of compensation and safety guarantees. In the neutral area, those against siting the facility in their region came to 94 percent, which is the same as the proportion of Ofalu respondents who opposed the current siting. In the Paks area there was less opposition but here, too, hardly anyone wanted it (5 percent were for it, 12 percent conditionally agreed). Respondents argued that their concerns were not merely associated with the direct dangers caused by the facility, but also with other disadvantages. In the Ofalu area 30 percent were afraid of outmigration from the area, 15 percent of a drop in tourism, and a further 10 percent of a decrease in property values.

### Attitudes Towards Experts

In the local communities' attitude towards the scientific arguments concerning radioactive waste disposal -- being laymen in the field -- the trust they placed in the competence and honesty of the experts was a determining factor.

We asked the interviewees, how they evaluated the competence and trustworthiness of the various groups of experts.  We asked how well the experts are able to judge whether radioactive waste could be safely disposed of in Ofalu, and how much they trusted in the honesty of the experts.  Four possible answers were given (4= completely, 3= more-or-less, 2= not very much, 1= absolutely not).  We asked about the following six groups:  the experts of the nuclear power plant, the independent experts brought in by the inhabitants, the councils' experts, the experts from the ministries, those from the Hungarian Academy of Sciences, and foreign experts (Table 10-5a/b).

In general, faith in the experts' trustworthiness was less than that in their competence.  In all regions, respondents evaluated both the competence and trustworthiness of the foreign experts as relatively high.  In the Paks and neutral regions, the experts of the Academy of Sciences were considered most competent and trustworthy, whereas the opinion of the Ofalu respondents was less positive about them.  In the Ofalu region, experts of the nuclear power plant were estimated as the least competent and trustworthy of all the expert groups, whereas they were less negatively evaluated in the other regions.

In the Ofalu region, the inhabitants' opinion of the different groups of experts was strongly related to their activities concerning the siting.  In general, their opinion about the experts was more negative than those of the other groups, except towards the independent experts and the council experts who were against the facility.

The prestige of the Hungarian Academy of Sciences is revealed in that in the Paks and neutral areas it was most favored both for its ability and integrity, followed closely by the foreign experts.  Trust in the Academy in the Ofalu area might have been similarly great at the beginning of the siting process, but by the time of this survey attitudes had changed.  The Academy had sent a committee to investigate the suitability of the site.  The fate of the facility was in their hands but the Academy did not live up to the expectations of the Ofalu inhabitants.  Its decision was postponed, but some of its opinions, unfavorable to the local community, leaked out, all of which raised doubts in people's minds.

Although nearly one-half of those in the Ofalu region, 70 percent in the Paks region, and 62 percent in the neutral region expressed the opinion that the Academy was capable of judging the safety of the facility, and an even greater number trusted its honesty, few said that if the Academy found Ofalu a suitable site that they would agree to the Ofalu siting (10 percent in the

Ofalu region, 36 percent in the Paks region, and 35 percent in the neutral region).

Table 10-5a. Estimation of the competence and trustworthiness of the experts

### Competence

| Expert | Ofalu | Paks | Neutr. | N | F | Sig. |
|---|---|---|---|---|---|---|
| Foreign | 3.06 | 3.29 | 2.98 | 692 | 3.86 | 0.02 |
| Independent | 2.85 | 2.71 | 2.45 | 692 | 2.67 | 0.06 |
| Academy | 2.77 | 3.32 | 3.04 | 691 | 13.7 | 0.001 |
| Ministry | 2.30 | 2.45 | 2.32 | 691 | 1.4 | --- |
| Council | 2.18 | 1.80 | 1.80 | 692 | 10.16 | 0.001 |
| Nuclear plant | 2.16 | 2.82 | 2.74 | 691 | 26.9 | 0.001 |

Table 10-5b. Estimation of the competence and trustworthiness of the experts

### Trustworthiness

| Expert | Ofalu | Paks | Neutr. | N | F | Sig. |
|---|---|---|---|---|---|---|
| Foreign | 2.87 | 3.03 | 2.70 | 688 | 2.66 | 0.07 |
| Independent | 2.86 | 2.50 | 2.28 | 687 | 7.69 | 0.001 |
| Academy | 2.47 | 3.10 | 2.73 | 690 | 10.97 | 0.001 |
| Ministry | 2.04 | 2.26 | 2.07 | 698 | 2.2 | --- |
| Council | 2.47 | 1.79 | 1.74 | 687 | 30.0 | 0.001 |
| Nuclear plant | 1.50 | 2.52 | 2.24 | 690 | 62.8 | 0.001 |

## The Preferred Means of Decision Making, Operation and Supervision

During the first ten years of the siting process, the communities around the proposed sites were completely excluded from the decision-making process. The organizers of the local opposition stressed that this had been instrumental in the formation of the opposition. The results of the survey, however, indicated that to the majority of respondents public involvement was of secondary importance.

Those questioned -- independent of the way they were to be affected -- thought the public had only a small role to play in the siting of the facility. Altogether only 10 percent felt that opinions of local residents should be taken into account in the decisions. Few felt that the nation should decide as a whole. Most of the respondents thought that the experts should choose a suitable site.

The most apparent difference between the variously affected groups was that a significant proportion of those living near the nuclear power station (29 percent) said it was necessary to take into account the national attitude to the decision, whereas only 8 percent in the Ofalu area, and only 13 percent in the neutral area expressed this opinion.

Residents of the three regions were asked about who should be involved in supervising the facility. Table 10-6 indicates that few respondents felt that local

Table 10-6. Who Should Supervise the Safeguarding of the Radioactive Waste Repository (percentages)[3]

|  | Ofalu | Paks | Neutral | N | F | Sig. |
|---|---|---|---|---|---|---|
| Independent experts | 35.9 | 36.0 | 30.8 | 498 | 2.29 | 0.1 |
| Nuclear power plant experts | 15.1 | 48.5 | 49.5 | 696 | 36.75 | 0.001 |
| International Atomic Energy Commission | 11.8 | 38.8 | 9.6 | 696 | 2.99 | 0.05 |
| Environmental protection organizations | 14.1 | 7.5 | 12.1 | 696 | 0.26 | --- |
| Local inhabitants | 11.1 | 9.5 | 5.6 | 696 | 1.98 | 0.1 |
| Councils | 8.4 | 1.5 | 2.0 | 696 | 0.67 | --- |
| Other | 14.5 | 3.0 | 14.6 | 693 | 1.3 | --- |

communities could competently supervise the facility. In the Ofalu region about 11 percent thought this was a task for the local inhabitants and 8 percent thought the councils should be included. Fourteen percent felt that organizations specialized in environmental protection should be involved whereas 12 percent would involve the International Atomic Energy Commission. Thirty-six percent of the respondents, however, thought supervising was a job for independent experts, often mentioned along with the nuclear power station's experts.

*Attitudes to Compensation*

The willingness to accept compensation for hosting the facility proportionally increased with the distance from the site: 27 percent in the Ofalu area, 39 percent in the Paks area, and 55 percent in the neutral area thought it was proper that the inhabitants receive compensation. A small proportion of the respondents thought of it in terms of financial compensation (6 - 8 percent of the responses), whereas 15 percent in the Ofalu region, 33 percent in the Paks region, and 41 percent in the neutral region thought in-kind compensation was more appropriate.

In all three regions about one-third of those who refused compensation gave as their reason that damages caused by the facility could not be compensated for. In the Ofalu region, two other types of arguments were brought up: first, the unconditional rejection of the facility ("We do not need compensation because they cannot build it here."), and second, moral reasons ("It is bribery," "We cannot be bought," "We will not give up our village".).

## Conclusions

The results of the public opinion survey confirmed that attitudes to the siting of a radioactive waste disposal facility are strongly related to how the individuals and the communities are affected by the risks, disadvantages, and benefits of the facility and the nuclear industry in general. Attitudes to the siting also seem related to the individual's knowledge about nuclear technologies, as well as environmental and political orientations. Finally, economic and social circumstances of the local communities also appear as important factors.

The research was done in an extraordinary historical situation when events mirrored the change in power politics. The rejection of the disposal facility in the given political situation signified an anti-government, anti-regime stance and occurred in this light. Therefore, on the basis of the results of the survey, making forecasts about the attitudes of the public in future siting controversies is difficult.

The results, however, illustrate the importance of eliciting public opinion over time. If the proponents of the siting had known the inhabitants' attitudes before the decisions had been made, they might have changed the direction of events. Public opinion research, combined with other models of public involvement, could serve as a basis for other similar decision processes in the future.

## Notes

1. These areas: inflation, environmental pollution, declining living standards, growing rate of crimes, decrease in population growth, sharpening political tensions, unemployment.

2. Source: Harsfalvi, T., Kulcsar, L., Misovecz T. (1989).

3. The respondents could choose one or more of the groups or organizations. Thus the total exceeds 100 percent.

4. This was an open question. The table is based on the first choices. Taking the second choices into consideration would modify the table because most put environmental and health safeguards as their second choice.

## References

Covello, V. 1983. "The Perception of Technological Risks: A Literature Review". *Technological Forecasting and Social Change*, 23, 285-297.

Dobossy, I., & Kulcsar, L. 1985. "Az okologiai tudat es viselkedes tarsadalmi tenyezoi. (Social factors of ecological consciousness and behavior)", *Szociologia*, 3-3.

Dunlap, E., & Baxter, K. 1988. Public Reaction to Siting a High-Level Nuclear Waste Repository at Hanford. SESRC, Washington.

Earle, T.C. & Lindell, M.K. 1982. Public Perception of Industrial Risks: A Free-Response Approach. Battelle Human Affairs Research Centers, Seattle, WA.

Freudenburg, W.R., & Rosa, E., eds. 1984. Public Reactions to Nuclear Power. AAAS Selected Symposium 93.

Harsfalvi, T., Kulcsar, L., & Misovecz T. 1989. Kornyezetgazdalkodas es lakossagi tudat. (Environmental Management and Population Awareness), Study, commissioned by the Ministry of Environment, Budapest, Hungary.

Johnson, B.B. 1987. Public Concerns and the Public Role in Siting Nuclear and Chemical Waste Facilities. *Environmental Management*, 11, 571-586.

Inglehart, 1990. Cultural Shift in Advanced Industrial Society. Princeton University Press, Princeton.

Morell, D. & Magorian, C. 1982. Siting Hazardous Facilities: Local Opposition and the Myth of Preemption. Ballinger, Cambridge, MA.

Szirmai, V. (this volume). "The Structural Mechanisms of the Organization of Ecological-Social Movements in Hungary".

Valko, E. 1989. Energiaugyek a kozvelemenyben (Public Opinion about Energy Issues), Study, Hungarian Institute for Public Opinion Research, Budapest, Hungary.

van der Pligt, J. (in press). Nuclear Energy and the Public. Blackwell, Oxford, UK.

Vari, A., & Farago, K. 1991. From Open Debate to Position War: Siting a Radioactive Waste Repository in Hungary. *Waste Management, 11.* 173-182.

Vari, A., Kemp, R., & Mumpower, J.L. 1991. Public Concerns about LLRW Facility Siting: A Comparative Study. *Journal of Cross-Cultural Psychology, 22,* No. 1. 83-102.

Chapter 11

# ECOLOGICAL CONSCIOUSNESS IN THE USSR: ENTERING THE 1990S

Boris Z. Doktorov, Boris M. Firsov, Viatcheslav V. Safronov

This paper is an attempt at summing up the results of research into the ecological consciousness in the USSR. The gathering of data, reflecting the attitude of the entire Soviet population towards ecological problems, became possible only in the late 1980s, when the All-Union Center for Study of Public Opinion was set up. The authors developed three ecological blocks for the Center and used them in the 1989 - 1990 opinion polls (April, 1989 representative sample of the urban population of the USSR, 1019 respondents; November 1989, representative sample of the USSR adult population, 1516 respondents; August 1990, representative sample of the entire USSR population, 2949 respondents). Data used in this paper are drawn from these national surveys.

## The Soviet People's Concern About Ecological Problems

*The Degree of Concern*

Even if we do not lay emphasis upon those cares and concerns caused objectively by socio-political conflicts in the country, such as aggravation of inter-ethnic relations and economic crisis, we should admit that most of the population faces numerous hardships and the citizens of the country must every day wage a bitter struggle for subsistence. No special proofs are needed here. But we would like to dwell upon the economic component. All the cells of mass consciousness are "stuffed" with economic matter. In such conditions public opinion will first reflect the socioeconomic situation, considering it the main social evil. Opinion polls, conducted by the All-Union Center for Study of Public Opinion, witness that in 1989 - 1990 low income of the people, price growth (inflation), shortage of food products and consumer goods, and housing problems caused the greatest concern among the respondents. Pollution of the environment ranked fifth in the list of the most acute problems which arouse society's concern, coming prior to such burning issues as bureaucracy, corruption, low standards of medical service, inter-ethnic conflicts, degradation of public morals, and the growth of strikes. All the measures of the authorities in the last two to two and a half years have brought forth and strengthened the awareness of mounting poverty. We dare to forecast that in the near future this feeling will dominate in the rank and file Soviet citizen.

*A. Vari and P. Tamas (eds.), Environment and Democratic Transition, 249–267.*
© 1993 *Kluwer Academic Publishers. Printed in the Netherlands.*

But our forecast does not mean that the ecological consciousness accepts the state of the environment. It is just the other way around: the public opinion polls testify that no less than 85 to 90 percent of the population are concerned and worried about the ecological situation in the country. Only a few respondents showed no concern and no more than five percent found it difficult to formulate their viewpoint. Therefore, ecological issues are in the center and not in the periphery of mass consciousness.

According to recent (August 1990) national survey data, approximately equal (and high) concern of the population exists about the state of the environment on all four levels: local, regional, nationwide and global (Table 11-1). The similarity of appraisal may be accounted for by two circumstances. We deal with a sort of generalization and extrapolation of environmental concern from a local level to a higher one. Also, the influence of mass media makes itself felt. Information provided by the media about what goes on in the region, in the country, or in the world becomes generalized.

Table 11-1. The Concern of the Soviet Population about the State of Environment on Different Levels of Locality (August, 1990, the USSR population, 2949 respondents, %)

| "To what extent are you concerned about the state of environment | Very Much Concerned | Rather Concerned | Rather Not Concerned | Not Concerned At All | Don't Know |
|---|---|---|---|---|---|
| ... in your town (settlement, village) and in its vicinity?" | 55 | 35 | 5 | 2 | 3 |
| ... in your region or republic?" | 57 | 34 | 3 | 1 | 5 |
| ... in our country?" | 60 | 30 | 3 | 1 | 6 |
| ... on our planet?" | 59 | 28 | 4 | 1 | 8 |

Soviet people are concerned about the development of ecological problems over a span of time. According to the April 1989 survey, almost 70 percent of USSR town population reported perceived degradation of the environment in their communities over the last 10 years. Only 10 percent of the respondents perceived a slight improvement. As displayed in Table 11-2, the forecasts are also quite pessimistic. Fewer than 20 percent of the respondents in the November 1989 and August 1990 surveys thought that the quality of their local environment would improve during the next 3-to-5 years.

Table 11-2. Estimation of the Changes in the Environment in the Nearest Future

| "How do you think the environment in your place of residence will change most probably in the next 3-5 years?" | 1989 November USSR Population (1516 resp.,%) | 1990 August USSR Population (2949 resp.,%) |
|---|---|---|
| Will improve considerably | 4 | 4 |
| Will improve slightly | 14 | 14 |
| Will not change | 24 | 29 |
| Will change for the worse slightly | 12 | 16 |
| Will considerably change for the worse | 12 | 14 |
| Don't know | 34 | 23 |

*Sources and Types of Concern*

The environment has undoubtedly been degraded for the last decades. First and foremost, public opinion has reacted to the most obvious dangers, the ones which catch the eye and can be perceived immediately: polluted air; dirty streets, deteriorated buildings; polluted ponds, lakes, and rivers; the unsanitary state of urban areas; and herbicides and pesticides in fruit and vegetables (Table 11-3). The contamination of drinking water, the decrease in the number of species of flora and fauna, the pollution of recreation places, and noise and radiation levels are regarded as less important factors. No more than one-quarter of those polled are concerned about them (Table 11-3). The urban population is even less worried about climatic change, the deterioration of soil, the lowering of its natural fertility, acid rain and about such other problems, whose understanding requires a broad educational background and knowledge (Table 11-3).

Table 11-3. The Concern of the USSR Urban Population About Particular Aspects of the Ecological Situation (April 1989, 1019 respondents, %)

| What particularly worries you in your town and its environs? (The respondents were allowed to select several items) | % |
|---|---|
| Air pollution | 55 |
| Dirty streets, deteriorating buildings | 45 |
| Polluted rivers, lakes, seas | 44 |
| Unsanitary conditions of urban territories -- rubbish, dumps | 38 |
| Noxious chemical substances in fruits and vegetables | 35 |
| Contaminated drinking water | 25 |
| The trampling down, damaging, pollution of parks, gardens, recreation places | 23 |
| Excessive level of noise | 18 |
| Excessive level of radiation | 15 |
| Absence of parks, gardens, green grounds | 14 |
| Vanishing of certain species of birds, fish, animals and plants | 14 |
| Changing climate | 13 |
| Deterioration of soil and its natural fertility | 12 |
| Ruining of historical monuments, desertion of old country estates, parks | 12 |
| Vanishing of woods, their unsatisfactory condition | 11 |
| Acid rains | 11 |
| Vermin and pests | 9 |
| Reduced levels of lakes and rivers, emergence of deserts, swamps and other changes in the landscape | 9 |
| Unfitness of water for sprinkling | 3 |
| Other factors | 1 |

Recently the attention of the people has more and more frequently begun to switch from the obvious to covert harm inflicted on people by poisonous admixtures filling the environment "imperceptibly" (Mitchell, 1989). For example, peoples' concerns about radiation have significantly increased since the Chernobyl disaster. This conclusion is backed up by the distribution of opinions with respect to nuclear power engineering. According to the survey of August 1990, only 13 percent of the Soviet population support the use of this technology in power production, whereas 54 percent reject it.

Recent surveys indicate that the Soviet people closely link the state of the environment with their health condition (Table 11-4). Most people would like to channel the state budget to health protection (59 percent), agriculture (39 percent), combatting crime (36 percent), social security (36 percent), and only a small minority would suggest to spend it for defense (10 percent) or aid to other countries (2 percent).

Table 11-4. Relationship Between Environment and Health

| Do you think the state of the environment in your place of residence is dangerous for health? | 1989 November USSR Population (1516 resp.,%) | 1990 August USSR Population (2949 resp.,%) |
| --- | --- | --- |
| Yes | 60 | 56 |
| No | 19 | 22 |
| Don't Know | 21 | 22 |

*General Types of Ecological Views*

Using the method of basic components analyzing the August 1990 survey data enables us to reveal a two-dimensional space containing the main types of ecological views. The first factor reflects how much the people are concerned about the existing ecological situation. This factor reflects various levels of concern (highly concerned - less concerned). The second factor reflects the attitude to potential danger of environmental pollution (optimism -- pessimism). Generally speaking, at least four polar types, situated in different quarters of the above-mentioned factor space exist. In the present case we shall single out only three generalized types of ecological views: they embrace about 80 percent of the sample.

Type 1: Ecologists (51 percent). This group consists of those who are greatly concerned about the environment in their locality, province or country, or in the world. Ecologists find it necessary to channel more means to environmental protection and, within their powers, are ready to finance it (see below). They demand immediate actions against further deterioration of the environment and do not intend to shut their eyes to negative consequences of technological progress for the sake of overcoming economic difficulties.

Type 2: Concerned Pessimists (16 percent). Representatives of this type are less worried about the current ecological situation than ecologists and a bit more pessimistic in their forecast.

Type 3: Concerned Optimists (12 percent). Representatives of this type do not differ from concerned pessimists by the degree of their concern, but stick to the opposing views in the estimation of future danger.

We have failed to detect considerable socio-demographic distinctions in the structure of the above types. The results show that public opinion of the Soviet population is characterized by a high degree of consensus with respect to a relatively high level of concern about environment. The uniformity of view must be accounted for rather than the distinctions between the majority and a certain minority: in other words, the reasons for the serious ecological concern of the people need to be ascertained.

## Analysis of Concerns and Explanations

*The Population's Readiness to Sacrifice for the Sake of the Environment*

To what extent is the population of a country ready to make certain personal sacrifices for the sake of protecting the environment? Many researchers have taken up this issue. It is logically linked with the scope of support which the population in different countries will be able to provide for various measures aimed at solving serious ecological problems.

The majority of the Soviet people are against overcoming the economic crisis at the cost of the environment. (Table 11-5) More so, almost 80 percent of those polled said that all available means should be used for protecting and preserving the environment. Appraising the real difficulties the economy has run into, however, many people (no less than 80 percent) have not fully acknowledged the conflict between economic growth and environmental protection. They believe the two crises should be overcome simultaneously.

The situation becomes more contradictory when the question about the sources of material means is posed. The state budget is ranked first. As for other means (personal material sacrifices for the sake of protecting nature) are concerned, the situation is more complicated. In the economic and ecological crisis, when the living standards go down continuously, almost each respondent is confronted with an extremely difficult choice. The respondents are between Scylla and Charybdis, i.e., caught between the low living standards and the noxious influence of the polluted environment on their health. Abiding by the experience and methods of international research, we have made an attempt at analyzing this contradictory situation with the aid of the aggregate data obtained by polling urban and rural populations of the Russian Federation.

Table 11-5.  Readiness to Back Up Measures for Protecting Environment
(August, 1990 the USSR Population, 2949 resp.,  percent)

| Judgments | Fully Agree | Somewhat Agree | Somewhat Disagree | Fully Disagree |
|---|---|---|---|---|
| One will have to put up with the pollution of environment to cope with the economic difficulties | 6 | 8 | 28 | 58 |
| The environment can be preserved only if we agree to lower our living standards | 7 | 14 | 38 | 41 |
| All the means must be channeled to the environment's protection | 41 | 36 | 18 | 5 |
| I am ready to transfer a certain sum of money monthly if I am sure the money will be allocated for the environment's protection | 28 | 33 | 20 | 19 |
| I shall agree to have taxes raised if I am absolutely sure the means will be channeled to the environment's protection | 22 | 31 | 26 | 21 |
| The State must find the means to protect the environment.  It is no concern of mine. | 29 | 23 | 33 | 15 |

Due to the method, only slightly different from the one used by Inglehart (Inglehart 1991) the index of "readiness for material sacrifices for protecting the environment" is calculated. Three types are singled out:

a) Ready for sacrifices, i.e., agree to transfer money to an ecological fund if sure the money will be well disposed of; on the same terms agree to have taxes raised; do not think that it is only the government which should allocate money for environment protection (33 percent).

b) Mixed type (12 percent),

c) Not ready to sacrifice, i.e., do not agree to transfer money and have taxes raised; they think that it is up to the government to find means (25 percent), and

d) Uncertain or no answer (30 percent)

The percentage of those who are eager to make material sacrifices for the sake of protection of the environment in Russia is compared with that of other countries (Inglehart, 1991) are shown in Table 11-6.

Table 11-6. Readiness to Make Material Sacrifices in Different Countries

| Country | % |
| --- | --- |
| Republic of Korea | 58 |
| Czechoslovakia | 53 |
| Mexico | 50 |
| Japan | 47 |
| Chile | 46 |
| Canada | 42 |
| United States | 40 |
| Spain | 34 |
| Russian Federation | 33 |
| Nigeria | 32 |

*Source*: Inglehart, 1991; and August 1990, The Population of the Russian Federation, 1607 resp.

The Russian data conform to the general picture. In general the concern of the people about ecological matters in rich countries is more acute than in poor ones.     Objective ecological situation, degree, and scope of environmental pollution (determined largely by economic and industrial growth rates, the area of the country, and the concentration of production), the efficiency of measures, and the duration of combatting the pollution.

The Russian Federation occupies vast territory, its economic growth rates have been relatively slow of late, and the problems of environmental pollution are quite pressing, but were concealed from the public until recently. All this has restrained the development of ecological consciousness. Another no less important factor is that the impact of the ecological situation is postponed in time. Despite the acuteness of the environmental problems, the consciousness of the population reacts to the economic crisis. Relatively few are concerned with the protection of the environment under the conditions of a food shortage and a mounting inflation.

These explanations are not exhaustive. The research into tendencies towards public opinion transformation with respect to environmental protection in highly developed countries, with considerably improving environmental conditions, does not attest to the flagging, but rather the growth, of the support offered by the people. Moreover, in the United States the ecological course has drawn significant support even in the period of economic crisis (the mid 1970s and early 1980s).

## Materialist and Postmaterialist Orientations

In recent explanations given for the deep concern of the public about the condition of the environment, social values play a very important role (Downs, 1972; Milbrath, 1984; Dalton, 1984; Dunlap, 1989; Inglehart, 1990). The premise is that a certain shift in the structure of basic priorities (including views of economic, social, political, and ecological issues) is being made in developed countries due to rapid economic growth. Materialist values, implying first of all economic and physical security, begin to be ousted by postmaterialist values because emphasis is laid upon quality of life, improvement of social and natural environment, and upon an individual's self-realization. (Inglehart 1990)

The shift towards postmaterialist values is an essential factor whose influence predetermines the public concern about the condition of the environment and the readiness to lend support to measures aimed at protecting it. It is the postmaterialists who are eager to vote for greens and join ecological social movements. Among other variables materialist/ postmaterialist orientations possess the greatest predictive power of participation (real or potential) in the ecological movement.

The polarization of views along the axis of materialist/postmaterialist values can be traced out not only in the West, but also in any part of the globe. The component analysis of value issues (six materialist and six postmaterialist positions), reveals distinctly a factor shared in common by the United States, Canada, the industrial countries of Western Europe and of Eastern Asia (Japan and Republic of Korea). Similar tendencies (though not so clear cut) are observed in factor solutions for Eastern European countries

(Poland, Czechoslovakia), Latin American countries (Chile, Mexico) and African countries (Nigeria) (Inglehart, 1991).

Do segments with materialist and postmaterialist orientations exist in contemporary Russia? Table 11-7 contains the questionnaire eliciting preferences of the Russian population towards materialist (judgments A and C) and postmaterialist (judgments B and D) values.

Table 11-7. Questionnaire Eliciting Value Preferences (August, 1990, The Population of the Russian Federation 1607 resp.)

The list below contains some of the goals of political activity. Which one would you find most important? Which goal would rank second? Third?

A) Maintaining order in the nation
B) Giving people more say in important government decisions
C) Fighting rising prices
D) Protecting freedom of speech

Table 11-8 (the bottom line) shows the distribution of the polled according to the index of materialist/postmaterialist values. The population of the Russian Federation has by far more materialists than adepts of postmaterialist orientations. The pure postmaterialists are few (five percent). Nevertheless, taking into account a considerable number of people who stick to mixed value orientations (with postmaterialist trends prevailing by 21 percent), it may be asserted that there exists a certain stratification of value orientations in Russia. But the degree of manifestation is low and it is in keeping with the assumption of the theory under study, i.e., that the prevalence of postmaterialist outlook depends on the society's economic well-being. For example, the ratio of materialists and postmaterialists in six developed Western European countries was 4:1 in 1970, whereas in 1989 it reduced to 4:3. In Russia the ratio was 9:1 in 1990.

To better grasp the correlation between values and age, one should dwell upon the conditions each age group underwent during its socialization. From the late 1940's until the early 1960's, certain economic growth had been observed in the USSR which entailed a relative growth of the peoples' well being. Then the economic growth gave way to stagnation. But in the period of stagnation the sense of economic security did not vanish. It was associated with guaranteed work, earnings, housing, free education, and medical care. A certain, although not high, saturation of the market with food products and consumer goods brought forth during the period of early socialization of those born in the first post-war decades explains a tendency to stick to

Table 11-8. Materialist/Postmaterialist Values in Various Age Groups
(August, 1990, the Population of the Russian Federation 1607 resp., %)

| Age Group | Index of Values | | | |
| | MM | MP | PM | PP |
| --- | --- | --- | --- | --- |
| Below 20 | 35 | 35 | 20 | 10 |
| 20 - 24 | 31 | 29 | 31 | 9 |
| 25 - 29 | 34 | 27 | 32 | 7 |
| 30 - 39 | 45 | 26 | 22 | 7 |
| 40 - 49 | 50 | 21 | 23 | 6 |
| 50 - 54 | 53 | 27 | 18 | 2 |
| 55 - 59 | 52 | 29 | 18 | 1 |
| 60 and over | 63 | 24 | 11 | 2 |
| Share of the type among the respondents | 47 | 27 | 21 | 5 |

MM: The group of materialists who have given preference to judgements A and C
PP: The group of postmaterialists who have chosen judgments B and D
MP and PM: The groups with mixed value orientations

postmaterialist values, whereas older age groups' representatives, who had confronted the severest hardships early in their childhood and youth were apt to abide by materialist values. The results of the analysis (Table 11-8) agree with this broader historical interpretation.

In summary, the position on the axis of materialism-postmaterialism is linked with the distinctions between the life orientations of different generations. As the discriminant analysis shows (Table 11-9, Part I.), age, representing values of different generations, most of all affects value stratification.

Such variables as income and education have also a certain weight (Table 11-9, Part I.). But income (the material condition of an individual) affects only the views of the people representing the pure materialist type and those of the mixed materialist dominant type. People with low income stick to the first type, whereas the ones with big income stick to the second one. Distinctions in education make themselves felt along the entire scale. The share of postmaterialists is directly proportionate to the rise of educational level. Value stratification is most clear cut between the people with higher education and all the rest. The ratio of the materialists and the postmaterialists is 2:1 in the first group, whereas it is no less than 7:1 in the second one.

Another factor, the interest in politics requires special comment (Table 11-9, Part I.). The present generations have evidently socialized under different spiritual and ideological conditions which tell on their world outlook. Indeed, if stratification of the Russian population's values were distinguished only by the differing macroeconomic living conditions of the generations, the disparity between the positions of young and old people would hardly be so deep. Among the young, the ratio between the materialists and the postmaterialists is 3:1, whereas in older age groups it skyrockets to 30:1 despite the fact that welfare growth rates were very low before finally giving way to stagnation. Value distinctions among the poor and the rich are quite insignificant. Inglehart (1991) have detected this tendency in East European countries and in Spain and has put this disparity between theory and practice down to the influence of the transition from totalitarianism to democracy.

Table 11-9. Discriminant Analysis: (I) Materialist-Postmaterialist Values Index Dependent on Socio-Demographic Variables. (II) Readiness to Material Sacrifices Index Dependent on Socio-Demographic Variables. (August 1990, the Population of the Russian Federation, 1607 resp.)

| Variables | I | | II | |
|---|---|---|---|---|
| | Value of Inclusion | Statistics of Exclusion | Value of Inclusion | Statistics of Exclusion |
| Sex | 3.94 | | 0.95 | |
| Age | | 23.89 | 4.06 | |
| Education | | 18.51 | 0.91 | |
| Nationality | 0.52 | | 2.03 | |
| Family Position | 0.63 | | | 6.25 |
| Size of Settlement | 1.82 | | | 10.85 |
| Duratrion of Residence | | 5.01 | 9.17 | |
| Main Occupation | | 4.72 | 3.37 | |
| Character of Activity | 0.11 | | 0.83 | |
| Branch of National Economy | 2.58 | | | 13.25 |
| Party | 1.04 | | 0.16 | |
| Per Capita Monthly Income | | 11.07 | | 4.09 |
| Religion, Confession | 0.27 | | 0.60 | |
| Interest in Politics | | 14.57 | | 11.76 |
| Materialist-Postmaterialist Values | | | | 12.11 |

Whereas society was undergoing emancipation from fear and ideological control, the people were gaining independence, which was converting the population particularly the young one into politically conscious people. Opportunities to develop interest in politics have arisen. This interpretation may be universal and may hold good for the Soviet population's emancipation.

## Ecological Views: Differentiation

During the last twenty years in which the public's ecological opinion has been studied in developed countries, researchers have repeatedly tried to explain why some social groups were more concerned about ecological problems than others. Putting forward hypotheses about the influence of socio-structural variables, they persistently tried to calculate the coordinates of proponents of environmental movements and organizations and those of their opponents. Despite the efforts taken, the social basis for acute concern about ecological problems was not discovered. Even when the method of multidimensional analysis was resorted to, no more than 10 to 15 percent of variation in ecological views could be accounted for by the influence of structural variables. (Van Liere and Dunlap 1980) The only assumption left was that the nature of these variations was psychological and was connected with the respondents' mentality and with their world outlook. In such a paradigm, the role of values remained unaltered: they were quite likely to produce common views concerning the environment.

This idea only now seeming simple and evident, was proved by Inglehart (1991). From his data, postmaterialist values in principle symbolized a special desire to work for the sake of saving the environment and to make personal sacrifices to attain this end. The degree of this desire correlated with the living standards reached by a particular country, considering the population of the country at large.

Does the Russian Federation abide by these world rules and to what extent does the measured stratification of values affect ecological views of the population?

In the upper part of Table 11-10 the relationship between value orientation and the readiness to make sacrifices for the environment is displayed. As can be expected, postmaterialists displayed a greater desire to offer material sacrifices for the sake of environment. The share of those who agree to sanifice their own money is 40 percent for postmaterialists and 29 percent for materialists. So values polarize people's ecological views. But the degree of polarization is not very high in comparison with other countries. Nowadays the value stratification is not so clear-cut among the population of Russia because the economic crisis makes itself felt, forcing the people to be especially concerned about the economy.

The next step of analysis was aimed at discovering and revealing other factors of the population's polarization in terms of their readiness to material sacrifices. Discriminant analysis of data revealed that such variables as employment in a certain sector of the national economy, place of residence, and interest in politics are the most important factors of the differences in ecological outlooks (Table 11-10, Part II). People engaged in industry, transport, and construction are less inclined to donate their own money to improvement of the environment (30 percent). Representatives of the service sector (trade, public catering, municipal economy) and also the respondents working in the field of culture, education, art, and science are more inclined to donate (40 percent). The share of the people who refused to make personal material sacrifices was smaller in this group (20 percent versus 30 percent in the industrial sector) (Table 11-10).

The detected distinctions are rooted in our everyday life. Workers of the industrial sector are likelier to put up with ecologically unfavorable working conditions. Having taken these conditions for the norm, they do not consider environmental pollution so serious a factor as to donate their own money to its protection. The second explanation proceeds from the notion of group, or departmental interests. Rigid ecological courses, such as those linked with technological renewal of production, abolishing the monopolies of ministries (departments), and closing ecologically harmful enterprises runs counter to the interests of the production sphere workers. Meeting ecological requirements amounts to diminishing their share in revenues, lowering their wages, the threat of unemployment, and necessity for retraining for many of them. Hence, the underestimation of the environment's pollution and reluctance to finance protection is deliberate.

The next factor, affecting ecological outlooks profoundly, is place of residence (Table 11-10). The population of cities backs up the idea of environmental protection more ardently than that residing in the country, or in small towns. Probably the influence of objective conditions such as quality of air, water, and town areas, which contain higher concentration of noxious substances than the countryside, is operative.

Table 11-10 also illustrates correlation between interest in politics and aims at environment protection. The keener the interest in politics is, the firmer is the intention to protect the environment. Here the influence of cognitive mobilization is traceable. Eagerness to finance ecological funds, consent to have taxes raised, as well as involvement in ecological movements, imply a deliberate political choice in favor of ecological security, based on knowledge and experience.

Table 11-10.   Readiness to Make Sacrifices for the Sake of Environment
(August, 1990, the Population of the Russian Federation 1067 resp. %)

| Factors | Ready | Mixed Type | Not Ready |
|---|---|---|---|
| Values | | | |
| Materialist | 29 | 38 | 32 |
| Materialist-Postmaterialist | 37 | 45 | 18 |
| Postmaterialist-Materialist | 36 | 48 | 14 |
| Postmaterialist | 40 | 38 | 18 |
| Employment | | | |
| Industrial | 30 | 40 | 30 |
| Service | 40 | 20 | 20 |
| Residence | | | |
| Moscow, Leningrad | 39 | 36 | 25 |
| Regional (territorial) centre, capital of autonomous republic | 35 | 40 | 25 |
| District Centre | 38 | 36 | 26 |
| Small Town | 20 | 55 | 25 |
| Village | 27 | 49 | 24 |
| Interest in Politics | | | |
| To what degree do you show interest in politics? | | | |
| To a very great degree | 47 | 32 | 21 |
| To a great degree | 36 | 41 | 23 |
| Neither great nor small | 35 | 41 | 24 |
| To a small degree | 32 | 48 | 20 |
| Not interested at all | 18 | 44 | 30 |

At the same time, the significance of the above variables should not be
overestimated.  Bearing in mind that the bank of empirical data about
Russian population's ecological outlook is quite scanty, it is premature to
draw conclusions as to what extent materialist/postmaterialist values
concepts mirror Russian reality.  But still, facts require thorough research
into cultural changes and their consequences for a society.

## About Green Movements In A Nutshell

To be in a position to grasp the peculiarities of green movements in the
USSR, a more general societal background of their evolution should be

considered. According to the 1989-1990 public opinion polls, a number of trends, typical of the population's social activity on the whole, can be outlined. First of all, people only slightly involved in social movements started do drop out. Second, the goals of social movements began to undergo noticeable transformations: politically and economically oriented movements advanced into the foreground rapidly. New parties began to emerge and become legitimate.

The results of the opinion poll conducted by the All Union Center for Studying Public Opinion in May 1990 mirror the USSR population's political preferences (Table 11-11). According to the results there are no data testifying to universally (not locally) mounting confidence in the social activity of the greens.

Table 11-11. Willingness to Vote for the Various Parties (May 1990)

| Party | The USSR Population 2294 resp., % | Population of the Russian Federation 1725 resp., % |
|---|---|---|
| Communist Party of the Soviet Union | 19 | 21 |
| Communist Party on a democratic basis | 11 | 13 |
| Labour Parties | 10 | 9 |
| Greens | 7 | 8 |
| Social-Democratic Parties | 7 | 7 |
| Independent Communist Parties in the Republics | 4 | 2 |
| National-Patriotic Parties | 2 | 0.5 |

The numeric strength of socio-political organizations, movements, and groups is also not very great. In the middle of 1990 about 1.2 percent placed themselves in the Popular Front, one percent identified themselves with various informal movements, and about one percent referred to themselves as part of ecological movements. According to appraisals of experts, the USSR population is apt to back ecological movements hinging on particular ecological problems (e.g., disasters). The second trend -- ideological -- is not gaining mass support. The population may be very slowly imbibing the quite abstract, human environmental ideas.

The trait of the contemporary ecological movements is their quite clearly manifested political bias. The organizations, sticking to different political outlooks, do not cooperate in the main cause -- protection of nature. Hence, difficulties in consolidating the ecological movements and the conflicts between them are significant. Contacts between the greens and scientific experts are also hard to establish. Ecological enthusiasts strive to influence decision themselves, whereas experts are frequently reluctant to shoulder responsibility, being afraid of arousing the wrath of public opinion, thus granting the authorities (and sometimes their colleagues) an opportunity to criticize them for not always correct conclusions and suggestions.

Another phenomenon is the growth of social tension influenced by economic activity. The population is angry about their children's chronic diseases, about meager social security payments, about scantiness of the means channeled to compensation and rehabilitation of health.

All over the world, researchers, politicians, and leaders of mass green movements suggest that the way out of many ecological deadlocks is chiefly linked with using the mass media to involve the population in ecological activity. They claim that communication with the public via mass media should be regarded as an important lever for working out the democratic green policy and its efficient implementation.

It is doubtful, however, that this advice and these views could be effective in the contemporary Soviet conditions. As our own research testified, one-third of the population place their confidence in the official ecological information, about 40 percent doubt their truthfulness and unbiased attitude, and 10-12 percent do not trust it. If some official information concerning the ecological situation does not match what the city dweller has, he will be inclined to rely on his own observations (47 percent of the polled stick to this position). Channels of informal communications (relatives, acquaintances, friends) rank second (41 percent). Then come articles and lectures by journalists (27 percent), lectures delivered by experts (17 percent), and finally, public talks and evidence furnished by informal groups, public movements, and green movements (nine percent). All the other sources are considered to be less credible. Most respondents claim that ecological information requires checking up and confirmation.

In this situation, where people have no confidence in official data and information, the authorities cannot count on deriving support from the public. The tide of emotions would neutralize and turn down even judicious and balanced suggestions whenever they are put forward by the top men. The reaction of the people in such cases can be unpredictable.

Thus, three loosely connected factors are observable in the relations between the people and the environment. The first one is the perception of ecological problems by the public. On the whole, it is characterized by concern, anxiety and uneasiness. The second one is aspiration to action, in particular to lending personal material support to environmental policy. Various circumstances, first of all economic, prevent, people from supporting environmental policy. The third factor is the green movements. In the current acute and explosive situation ecological activists stand some chance to enlist the people's support. But as soon as the ecological conflict is at least partly resolved, the greens may be left without mass scale support. Protest bias and extremism of the ecological movements hamper to establish close contacts with science and with new political structures. The economic crisis reduces the space for the greens activity.

Today we are not in a position to outline the long term prospects of ecological movements in this country. The macro environment is too unstable. The short term prospects of the green movements do not seem bright.

# References

Cotgrove, S. F. & Duff, A. 1981. Environmental Values and Social Change. *British Journal of Sociology, 32* (I), pp. 92-110.

Dunlap, R. E. (Ed.). 1980. Ecology and the Social Sciences> An Emerging Paradigm. *The American Behavioral Scientist* (Special issue), 24 (I).

Dunlap, R. E. 1989. Public Opinion and Environmental Policy. In James P. Lester (Ed.): *Environmental Politics and Policy: Theories and Evidence.* Durham and London: Duke University Press. Pp. 89-134.

Dulton, R. J. 1984. *Environmentalism and Values Change in Western Democracies.* Paper prepared for delivery at the 1984 Annual Meeting of the American Political Science Association, the Washington Hilton, August 30 - September 2, 1984.

Firsov, B. M., Matulionis, A. & Tamoshunene R. (Eds.). 1987. *Problemy formirovaniya sovremennoy ekologicheskoy kultury* (The Problems of Modern Ecological Culture Formation). Vilnius: Institut filosofii, sotsiologii i prava.

Firsov, B. M. (Ed.). 1990. *Razrabotka nauchnyh osnov izucheniya i formirovaniya ekologicheskogo soznaniya naseleniya strany* (The Elaboration of Scientific Bases of the Study and Formation of the Ecological Consciousness of the Nation's Population). Moscow: Institut sotsiologii. Issue 9. Parts I-2.

Firsov, B. M. (Ed.). 1991. *Razrabotka nauchnyh osnov izucheniya i formirovaniya ekologicheskogo soznaniya naseleniya strany* (The Elaboration of Scientific Bases of the Study and Formation of the Ecological Consciousness of the Nation's Population). Moscow: Institut sotsiologii. Parts I-2.

Inglehart, R. 1990. *Culture Shift in Advanced Industrial Society.* Princeton, New Jersey: Princeton University Press.

Inglehart, R. 1991. *Postmaterialism and Environmentalism: The Human Component of Global Change.* Paper presented at the Annual Meeting of the American Association for the Advancement of Science. Washington, D.C., February 17-20, 1991.

Lauristin, M. & Firsov, B. M. (Eds.). 1987. *Massovaya kommuninikatsiya i ohrana okruzhauschey sredy* (Mass communication and the Protection of Environment). Tallinn: Eesti Raamat.

Milbrath, L. W. 1984. *Environmentalists: Vanguard for a New Society.* Albany, N.Y.: State University of New York Press.

Milbrath, L. W. 1986. Environmental Beliefs and Values. In Margaret G. Herman (Ed.): *Political Psychology.* San-Francisco, CA: Jossey-Bass.

Mitchell, R. C. 1984. Public Opinion and Environmental Politics in the 1970s and 1980s. In N. J. Vig & M. E. Kraft (Eds.): *Environmental Policy in the 1980s: Reagan's New Agenda.* Washington, D.C.: Congressional Quaterly Press. Pp. 51-74.

Mitchell, R. C. 1989. From Conservation to Environmental Movement: The Development of the Modern Environment Lobbies. In M. J. Lacey (Ed.): *Government and Environmental Politics: Essays on Historical Development since World War Two.* Washington D.C.: Wilson Centre Press. Pp. 81-113.

Van Liere, K. D. & Dunlap, R. E. 1980. The Social Bases of Environmental Concern: A Review of Hypotheses, Explanations and Empirical Evidence. *The Public Opinion Quarterly, 44* (2) Pp. 181-197.

Chapter 12

# ENVIRONMENTAL RISKS IN A SOCIETY IN TRANSITION: PERCEPTIONS AND REACTIONS

Nikolai Genov

It was the environmentalist movement which signaled the coming changes in Bulgaria at the end of the eighties. Concern was provoked by the government's inability to handle the problems with the transboundary pollution of the town of Ruse on the Danube River. The nervous reaction of the authorities, who dissolved the citizens' national committee that had been established for protection of the town, was symptomatic of the fragility of the regime. Another overreaction was the police crackdown on a modest rally against hydrological projects held during the European Conference on Environment, which took place in Sofia in October 1989. Thus, the environmentalists turned out to be the only more-or-less organized collective actor that dared to challenge the regime before the beginning of the fundamental changes, only a fortnight later.

After the turn to democratic political institutions, the destruction of the environment became a battle cry. The veil of secrecy covering the level of environmental pollution was lifted. The shock of the public was predictable. Doubts and assumptions concerning negative effects of the rapid industrialization were largely confirmed. One could learn that the soil in vast areas around Sofia, Bourgas, Plovdiv, Kirdzhali and other towns was contaminated with chemicals and heavy metals, or that the concentration of sulfuric dioxide, lead aerosols and dust in the air was far above admissible levels in a number of industrial centers. An awareness of the fact that non-renewable resources had been overexploited in a shortsighted and ineffective manner was no less depressing. One could hardly imagine that the deposited industrial and household wastes, if dispersed evenly, would cover the country's territory with a layer 4.3 cm thick. Moreover, it became public that the large waste deposits were a time-bomb since they were not properly isolated and were likely to contaminate the underground waters for centuries (Yearbook, 1991; Environment and Development, 1992).

In the turmoil of 1990 and 1991, it seemed that environmental issues were destined to become a major topic on the political agenda. Moreover, they were regarded as a potential detonator of uncontrolled social unrest. The mass media was dominated by strikes, picketings and rallies demanding an immediate improvement of the environmental situation in settlements or enterprises. In fact, most of the protests were politically motivated. The environmental movements and, later the Green Party, seemed to develop

A. Vari and P. Tamas (eds.), Environment and Democratic Transition, 268–280.

into respectable political forces able to raise mass support. In some cases, such as the ban on the country's second nuclear power plant project, the actions of environmentalists succeeded; but, despite the then fashionable pressure to close the worst industrial polluters, the effects were rather modest. It was clear that the outcome of a radical closure of production lines would be a drastically increased level of unemployment.

About three years after the sweeping changes began, the picture is rather different. Environmental issues nearly disappeared from the political agenda. Environmentalists are no longer represented in the Parliament. Facing a 40 percent drop in the industrial production, a 50 percent decline in the purchasing power and half a million unemployed, the average Bulgarian seems to have become insensitive to environmental problems. Is this really the case? What are the trends in the mass perception of environmental risks? How are the reactions to environmental risks embedded in the value-normative and institutional contexts of the transition to a market economy and democratic political institutions? Some answers can be obtained from the data gathered in two consecutive years by a team from the Bulgarian Academy of Sciences, headed by the present author. The field studies were carried out in the capital city of Sofia with 1.2 million inhabitants. The samples covering the adult population consisted of N=1,630 in March 1991 and of N=1,533 in April 1992 and were based on electoral lists. The survey was conducted as home interviews. Another set of data comes from a nationwide survey of 1,700 persons carried out simultaneously with the survey in Sofia in April 1992. The tools, methods and sampling procedures were the same in both surveys.

*Risks, Fears and Mistrust of Institutions*

There are clear indications that the deepening economic crisis lead Bulgarians to perceive the threat of environmental risks less intensely. Comparing data obtained from identical questions in Sofia in 1991 and 1992, one may recognize the decline of social and psychological tensions directly caused by environmental issues, such as air pollution in the place of residence.

The reason for the above trend in the perception of environmental risks is not the economic crisis alone, however. It is true, basic economic needs have turned increasingly dominated mass consciousness. In April 1992, 77.4 percent of the adult population of Sofia regarded the state of the national economy as a very serious problem. But the weakening of the administrative and cultural control produced the outcome that an even larger segment of the interviewed persons, namely, 82.9 percent of them, saw crime as a very serious problem as well. Under these circumstances, it is not surprising at all that the state of the environment was regarded as a very serious problem by 52 percent. Moreover, a mere 2.2 percent of the adult population in Sofia indicated that the state of the environment was the most pressing problem of the country, while 47.5 percent considered the state of the economy to be the most pressing issue. No doubt, the result is partly due to the fact that the dramatic decline in industrial production contributed to the improvement of the environmental situation. Nevertheless, it would be wrong to assume that the concerns about environmental risks have nearly disappeared. This is neither the case in Sofia, nor in the country as a whole. Some data from the nationwide survey can provide the evidence supporting this claim.

The intensity of perceived risks country wide and in the capital is substantially different due to the high concentration of industry and means of transportation in Sofia. Another factor determining the difference is the higher educational level of the voters in the capital. As has been established by a number of surveys, there is a positive correlation between the level of education and the sensitivity to environmental risks.

What are the supposed causes or reasons for the risks in the area of residence? Figure 12-1 focuses on the contribution of industry, transportation and households to the environmental problems in the settlements, according to the assessment of the residents.

To illustrate the substantial differences between country-wide and Sofia data, 55.5 percent of the sample in Sofia consider that the transportation system causes serious or very serious environmental problems in the area of residence, while this applies to 31.5 percent of the country-wide sample. This is only one specific point indicating the range of substantial social differentiation's along the line of the access to the healthy environment as a typical common good.

The differences mark settlements, but also areas in the settlements themselves as shown in Figures 12-2 and 12-3. The awareness of the high

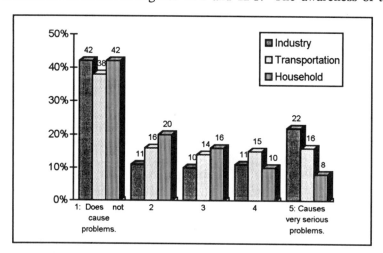

Figure 12-1. Factors causing environmental problems in the place of residence (countrywide).

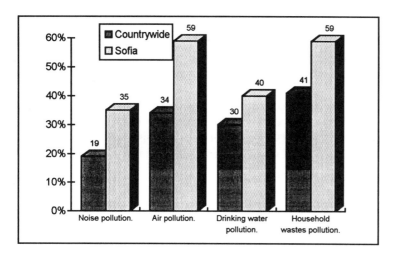

Figure 12-2. Noise pollution, air pollution, pollution of the drinking water and pollution from household wastes as serious and very serious problems in the place of residence countrywide and in Sofia.

intensity of exposure to environmental risks causes feelings of deprivation, dissatisfaction and fear. Of the nationwide sample, 30.1 percent state that they are strongly or very strongly afraid of the detrimental health effects which might be caused by residential environmental pollution. The higher

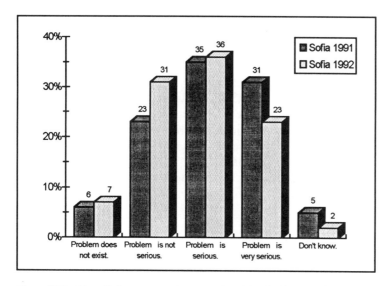

Figure 12-3. Air pollution as an environmental problem in the area of residence.

the level of declared knowledge about environmental risks, the higher the intensity of fears among persons who are exposed to them (coefficient of statistical correlation of Cramer, V = 0.189).

There are clear-cut differentiations concerning the environmental risks at the workplace as well. Figure 12-4 presents some assessments of work conditions.

Once more, the data show how deep the social differentiation based on environmental risks is. Of the 926 employees interviewed, 35 percent indicated that the noise is a serious or very serious problem in their work place. The same holds true for 38.6 percent concerning air pollution. The

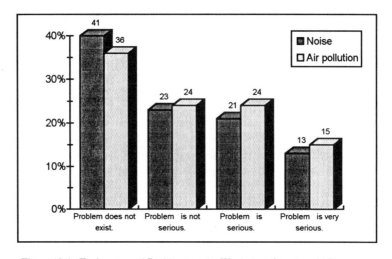

Figure 12-4. Environmental Problems at the Workplace (countrywide).

exposure of the main occupational groups to both types of environmental risks is quite different. The statistical correlation of occupational groups and the exposure to loud noise is V = .181, and for the risk of air pollution V = .182. According to expectations, the blue-collar workers are strongly over-represented among the people who are exposed to the above environmental risks at the work place.

What are the causes or reasons for environmental risks, which also include vibrations and radiation in the study under scrutiny? Figure 12-5 provides some clues in this respect.

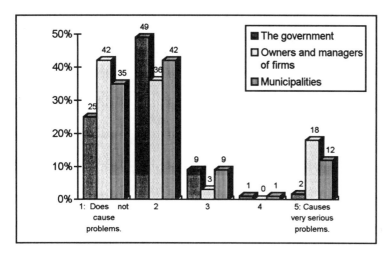

Figure 12-5. Causes of Environmental Risks at the Workplace (countrywide).

Another point is that the declared high, medium or low level of information about the environmental risks at the work place is a strong predictor of the declared intensity of fears ($V = .241$). The declared level of information is also an effective predictor of various kinds of activities such as participation in petitions or strikes, which aim at diminishing the environmental risks at the work place. In 1992, 10.3 percent of the interviewed employed persons from the nationwide sample stated that they have signed petitions, and 5.9 percent that they have participated in strikes provoked by environmental issues.

Thus, the sensitivity to environmental risks in the place of residence or at the work place cannot be eliminated by circumstances like looming unemployment, decline of purchasing power, political instability, and cultural disintegration. But the intensity of perception of environmental risks might decline under the extreme circumstances that are present in Bulgaria during the current transitional period. Since the risks threaten existential values, they might be easily revitalized and used by pressure groups, such as trade unions, political parties or cultural movements, as happens on various occasions.

This is all the more possible because together with the social and psychological pressures provoked by the environmental risks themselves, there is another important reason for tensions and conflicts. It is the deeply rooted conviction that, despite the existential importance of the environmental risks, the institutions which are directly responsible for managing them, are not doing what is necessary and expected in order to prevent the risks or to alleviate their negative effects. Figure 12-6 clarifies the point.

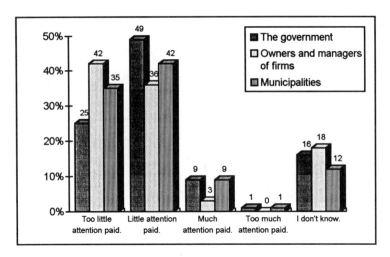

Figure 12-6. Attention Paid to the Environment by the Government, Owners, and Managers of Firms and Municipalities (countrywide).

The message cannot be misunderstood. A vast majority of the Bulgarian population is basically dissatisfied with the way environmental issues are handled by major social institutions. No doubt, what lurks behind the dissatisfaction is very often the traditional mistrust of politics and politicians. But there are also reasons which could only be explained by referring to current economic, political and cultural tensions and conflicts.

*Trends in the Relationship between Common Good and Private Interests*

In most general terms, there is a fundamental problem behind all the tensions and inconsistencies in public opinion concerning environmental issues in present-day Bulgaria. The core of the problem is located in the relationships between individual interests, initiative and responsibility, on the one hand, and the institutionalized control of the common good, on the other. Together with basic education and medical care, the provision of public order, etc., the environment is a typical common good in modern societies. This means that a healthy environment is expected to be accessible to everybody in principle and that institutionalized authorities are in charge of controlling the access to the common good.

The value-normative content of the institutionalized control of common good might be quite different, however. During the last decades, the management and control of the common good were defined in collectivist terms in Bulgaria. More precisely, individual initiative and responsibility concerning the environment had to comply with the principles of "Institutionalized Collectivism".

At first glance, this kind of value-normative orientation is optimal for the management of the environment as a common good.   However, the institutional pattern of centralized management of production, distribution and consumption based on state property and tight political control led to mismanagement in the long run.   One of the main reasons for this development is the suppression of the individual initiative and responsibility concerning the common good by the over centralized politics and economy. So the historical experience indicates that new value orientations and institutional patterns are needed in order to make the management of the common good more effective (Genov, 1992).

The need to institutionalize basically new value orientations is not limited to Bulgaria or to Eastern Europe.   Following some influential views (Inglehart, 1990), it is now common to analyze social trends in the most advanced countries in terms of the transition from materialist to post-materialist value orientations of the succeeding generations.  The transition is properly called "the silent revolution" which certainly has its internal tensions and conflicts caused by the persistent tradition of work organization, family life or leisure.   But they are quite different from the tensions and conflicts marking the current transitional period in Eastern Europe.  The effects are also different.   For instance, instead of the intergenerational differentiation of cultural patterns which is typical for Western Europe and North America, the concentration on the provision of basic necessities has narrowed the value-normative intergenerational differences in Bulgaria to the extent that they have become statistically irrelevant.   This holds true not only for environmental issues, but also for broader social problems.  The generational difference in the positive or negative opinions about the development of nuclear power production in the country is measured by the coefficient of statistical correlation of Cramer, $V = 0.120$.  In the acceptance or rejection of market economy as a favorable condition for solving environmental problems, the value is $V = 0.116$.   Even concerning the crucial political division between left, center and right, the generational variable shows a correlation of a mere $V = 0.123$.

In spite of the homogenizing effects of the current precarious conditions, some trends in attitudes towards environmental issues are recognizable.  Of the interviewed persons from the national sample, 48.9 percent indicated their willingness to pay a new "green" tax to be used for improving the environmental situation of the respective settlement. This willingness is quite surprising in light of the drop in the purchasing power of households. Moreover, not only is the present financial situation of the households grim, but prospects are not promising.  Of those interviewed country wide, 28.2 percent expect their financial situation to get even worse during the next one or two years, while 25.1 percent expect an improvement.  So the willingness to pay a new tax may reflect an emerging awareness that, without a personal involvement, the common good of the environment might be irreparably

destroyed. In addition, one may assume a positive correlation between the
level of the above awareness and the level of education. In fact, the level of
education, which is typically a weak predictor in the survey, turns out to be
quite relevant in predicting the willingness to pay a "green" tax.

Table 12-1. Level of Education and Readiness to Pay a "Green" Tax (country
wide)

|  | Payment of Tax | | |
| Education | No (Percent) | Yes (Percent) | Total (Percent) |
|---|---|---|---|
| No education and primary (4 cllasses) | 72.6 | 27.4 | 100.0 |
| Basic (7-8 classes) | 58.5 | 41.5 | 100.0 |
| Secondary general (11 classes) | 45.1 | 54.9 | 100.0 |
| Secondary special (technical etc., 11-12 classes) | 44.1 | 55.9 | 100.0 |
| Semi-higher (12-13 classes) | 34.1 | 65.9 | 100.0 |
| Higher (university, etc.) | 35.5 | 64.5 | 100.0 |

$$X^2_e = 87.65 \qquad\qquad V = 0.228$$

Some clarifications are needed. What changes along with the increase of
educational level? Is it the awareness that the individual contribution is vital
for solving burning social problems? Or, maybe it is the awareness that only
financially strong institutions could cope with the problems and that is why
they have to be supported. To put it in other terms, what matters: the
initiative and responsibility of individuals or the strength of institutions?
Clues for the proper answer might be provided by a comparison of data from
the nationwide study with data from the study on voters in Sofia. The crucial
point here is the experience that ideas and institutional arrangements taking
ground in the capital gradually gain influence all over the country.

In Sofia, 55.6 percent of respondents are willing to pay a new "green" tax,
which is higher than the overall country average. This is the outcome of the
influence of a complexity of factors. One of them is the concentration in the
capital of a higher proportion of younger and better educated people having
a broader horizon of interests than the average in the country. That is why
the trend towards individualization of initiative and responsibility that is
gaining influence in Sofia is quite indicative of current processes country

wide. A typical example is the changing attitudes towards the financial arrangements concerning education. Following the well-established national tradition, voters country wide lay the stress on the institutional solution of the problem. On the contrary, the voters in Sofia are definitely willing to shift the initiative and responsibility for obtaining education to the individual. Figure 12-7 clearly shows the shift.

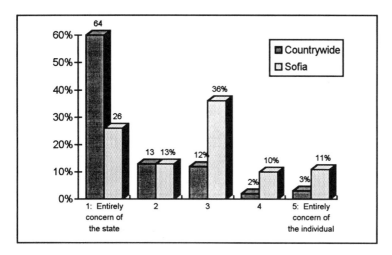

Figure 12-7. Whose Financial Concern Should be the Provision of Education?
(Countrywide and Sofia, 1992).

What appears from the above graphic is a clear trend to supplement, if not to replace, the institutionalized collectivism by *institutionalized individualism*. Preferences, decisions and actions of individuals are going to be more and more relevant for the management of the common good and, more precisely, for the protection of the environment. In fact, the trend is essential for the establishment and functioning of democratic institutions. They are supposed to provide the social space needed for personal initiative and responsibility, thus making the individual the pivot of institutional integration and change.

However, in the specific Bulgarian case, it is highly relevant to ask about the extent to which the public mind is ripe for accepting and supporting this qualitatively new relationship between individual interests and common good. The data from the study on environmental risks support the moderate optimism and raise substantial doubts in this respect. The reason for doubts might well be illustrated by the public assessment of the sweeping changes in the country after 1989. The fall of the *ancien regime* raised expectations enormously. The processes thereafter, however, brought about a high level of disenchantment.

Table 12-2. What Kind of Changes Took Place in Bulgarian Society During the Last One or Two Years?

|  | Countrywide | Sofia |
|---|---|---|
| The negative predominate | 31.1 | 25.2 |
| Positive and negative alike | 39.8 | 44.9 |
| The positive predominate | 14.2 | 20.7 |
| I don't know | 14.6 | 9.0 |
| NA | 0.3 | 0.2 |
| Total | 100.0 | 100.0 |

It would be naive to search for a simple explanation for the widely spread dissatisfaction with the transition to a market economy and democratic political institutions. The economic crisis and the political instability are certainly the main factors of uncertainty and anxiety which mark the current context of perception and reaction to environmental risks in the country. However, the value-normative and institutional transition from institutionalized collectivism to institutionalized individualism also has inherent tensions which contribute to the complexity of the current problem situation. It is the outcome of efforts to dismantle authoritarianism and to establish a democratic political order based on economic liberalism and cultural pluralism. The aggregated preferences to four political goals articulated by Ronald Inglehart are indicative of recent conflicts, but also of possible alternative developments in the future.

The above data are relevant as seen from two angles in the present context. First, they reveal a strong predomination of materialist over post-materialist value orientations both country wide and in Sofia. Since environmental issues are typically mixed with problems of quality of life and, thus, with post-materialist values, this is a clear signal that environmental policies cannot be in the very center of public attention under the present conditions. Second, a call for calm and order would receive very strong support. Having in mind the historical experience indicating that the call usually comes with the offer of a strong hand, which would establish and maintain calm and order, the prospects for a smooth democratic development are not necessarily bright in the country. One might be on firm ground speculating about the possible use of environmental issues as arguments for a turn to some kind of non-democratic form of government. Since they usually refer to the collective interest in order to establish their legitimacy, future environmental policies in the country may be guided by institutionalized collectivism anew.

Table 12-3. Four Goals Before the Country. Which is the Most Important One at Present?

|  | Countrywide | Sofia |
| --- | --- | --- |
| Calm and order in the Country | 56.3 | 60.3 |
| Impact of people on government's decisions | 5.0 | 7.6 |
| Fight to rising prices | 25.0 | 13.8 |
| Right of people to freely express their opinion | 9.9 | 14.7 |
| NA | 3.8 | 3.8 |
| Total | 100.0 | 100.0 |

The above data are relevant as seen from two angles in the present context. First, they reveal a strong predomination of materialist over post-materialist value orientations both country wide and in Sofia. Since environmental issues are typically mixed with problems of quality of life and, thus, with post-materialist values, this is a clear signal that environmental policies cannot be in the very center of public attention under the present conditions. Second, a call for calm and order would receive very strong support. Having in mind the historical experience indicating that the call usually comes with the offer of a strong hand, which would establish and maintain calm and order, the prospects for a smooth democratic development are not necessarily bright in the country. One might be on firm ground speculating about the possible use of environmental issues as arguments for a turn to some kind of non-democratic form of government. Since they usually refer to the collective interest in order to establish their legitimacy, future environmental policies in the country may be guided by institutionalized collectivism anew.

In the context of relatively stable advanced societies, it is tempting to discuss risks under the topic of the deviation from normalcy (Luhmann, 1991: 2f). In the Bulgarian society, which is going through a fundamental change towards a new type of social organization, risks are the normalcy. Values, norms and institutions are in flux. The chaos might prove to be highly constructive. But it might be destructive as well. Under such conditions environmental risks can motivate well-designed and implemented collective action aiming at rational institutional arrangements. But they can also lead to further weakening of social norms or to deepening of inequalities concerning the access to a safe environment as common good. It is the task of social science to describe and analyze the options. It is the task of politics to make the choice among options. One may only hope that the efforts of scientists, administrators and politicians will mobilize the intellectual and material resources which are needed for moving the country towards the trajectory of sustainable management of environmental risks.

# References

Environment And Development of the Republic of Bulgaria (1992). A National Report to the United Nations' Conference on the Environment and Development Brazil, '92. Sofia: Ministry of the Environment (in Bulgarian).

Genov, Nikolai (1992). "Environmental Risks and Public Policy: Trends in Bulgaria". In: Abdellatif Benachenhou, (ed.), *Environment and Development: Problems and Prospects for Sustainable Development.* Paris: UNESCO, pp.72-81.

Inglehart, Ronald (1990). *Culture Shift in Advanced Industrial Society.* Princeton: Princeton University Press.

Luhmann, Niklas (1991). *Soziologie des Risikos.* Berlin and New York: Walter de Gruyter.

Yearbook on the State of the Environment in the Republic of Bulgaria (1991). Sofia: Ministry of the Environment (in Bulgarian).

# INDEX